Topical Dosing: Beyond Pills & Needles

Topical Dosing: Beyond Pills & Needles

Marlon

A constant presence through life's twists

Table of contents

4 Introduction

The skin is the biggest human organ and accounts for 10 % of the body weight. With approximately 2 m² of surface area, the skin covers the entire body like an envelope and serves as the body's primary defense [1]. Owing to its complex multilayered structure, the skin possesses mechanical, regulatory, and sensation functions. The mechanical function provides both, a morphological flexible support apparatus with its connective and supporting tissue. It also serves as a mechanical barrier, protecting the body from irradiation exposure and exogenous environmental stresses and therefore forming a barrier that ensures decreased permeability for endogenous substrates and controlled permeability for primary gases and water loss. Additionally, the skin is an optimized barrier against exogenous substances like environmental pollutants and microorganisms. As an organ its regulatory functions include the control of oxidants, microbiome, immune response, metabolite synthesis and body temperature via sweat and water loss. The skin contains an extensive network of nerve cells which respond to environmental changes with a sensory function and is sensitive to heat, cold, touch, and pain.

To ensure site-specific penetration of an active pharmaceutical ingredient (API), a fundamental knowledge of skin morphology and physiology is required along with an understanding of its influence on API permeability. Because the skin is a complex and dynamic system, investigating these influences is indispensable for understanding and improving permeation and API delivery. [2, 3]

Until recent years, the skin has been primarily regarded as a protective cover of the body. Due to the difficulties overcoming this barrier, only a limited number of drugs are administered transdermally [4]. Exploitation of this route of administration is particularly important for therapeutic, diagnostic, prophylactic, and cosmetic applications [5]. Delivery of drugs in close proximity to the afflicted site inside the skin has advantages in terms of clinical treatment of disorders such as acne and cutaneous inflammatory diseases, wound healing, and inflammation [2, 6, 7]. Dependent on their mechanism of action, topically applied active ingredients have varying target-sites. Dermal drug delivery focuses on the site-specific localization of active ingredients in the skin to ensure controlled or prolonged drug delivery and systemic drug exposure. Site-specific drug delivery and a localized target in various skin layers adds complexity to the development of dermal drug delivery systems. Transdermal therapy aims at delivery to the systemic blood circulation and is an alternative to oral and intravenous drug delivery, if a significant concentration of API is ensured. [8, 9]

4.1 Morphology of the skin

Progressing inward, the human skin consists of three main layers: the epidermis, dermis, and subcutis.

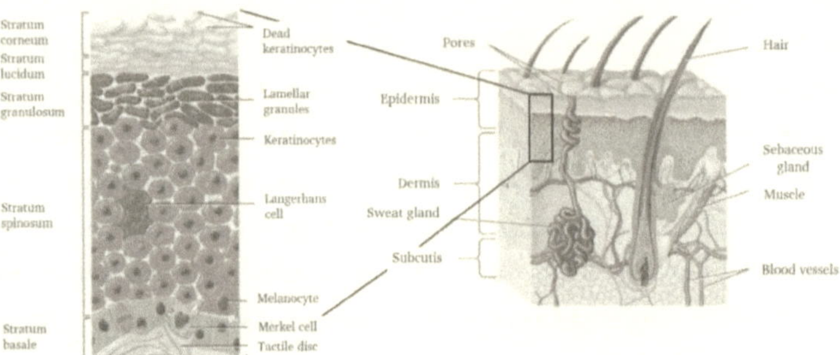

Figure 1: Morphology of the skin with its epidermis, dermis, and subcutis layer. [10]

Epidermis

The epidermis with its self-renewing properties consists of four distinctive layers and is divided into the viable epidermis and a non-viable dead horn layer (stratum corneum (SC)). Keratinocytes account for 95 % of the main cell type of the epidermis. These cells differentiate to form multiple layers [11, 12]. Keratinocytes migrate from the stratum basale to the stratum spinosum and stratum granulosum layer, which make up the viable epidermis, culminating into the final differentiation product, the SC; this process takes approximately 30 days [13]. The epidermis derives from the surface ectoderm, which becomes the embryonic epidermal basal single layer after epidermal commitment [14]. Thus, a single layer of ectoderm is formed at E9.5, and at E15.5 the proliferation is almost completely confined to the basal layer [15]. During this stage, a protective layer is already excised, the periderm, which protects the developing skin from constant exposure and from bacterial and environmental insults [15]. After the epidermis has fully stratified in the first few weeks and its differentiation into the five strata is completed, the protective periderm is shed [15, 16].

The multilayered epithelium contains a proliferating basal layer and a suprabasal layer in which the keratinocytes differentiate but no longer have the proliferation function [17]. The stratum basale is a single cell matrix layer that is attached to the basement membrane via hemidesmosomes and is considered a stem cell layer. Desmosomes are intercellular junctions of epithelia specialized for strong adhesion directly from cell to cell, in contrast to hemidesmosomes, which form adhesions between cells and the basement membrane [18]. In the basal layer, the epidermal stem cells proliferate, commit to differentiation, and migrate towards the environmental interface of the epidermis. During this stage of differentiation, the keratinocyte changes its horizontal axis formation to a volume richer, vertically, polygonal form, the stratum spinosum stage. This layer is formed by two to five prickle cell layers attached via desmosomes [13]. During this stage, the early differentiation marker proteins such as Keratin 1 and 10 are expressed by the keratinocytes [3]. The stratum granulosum contains one to three grain layers. In this stage of differentiation, the cells flatten out and the keratinocytes contain a high number of keratohyalin and membranous lamellar granules, referred to as lamellar bodies. These lamellar bodies, which are assembled in the Golgi, are tubular or ovoid-shaped, membrane-bound organelles. They synthesize and store glucosylceramides, cholesterin, free fatty acids, and phospholipids like sphingomyelin, which are precursors of SC lipids [19, 20]. Additionally, the lamellar bodies are responsible for the packing of antimicrobial peptides, proteins, protein-degrading enzymes and protease inhibitors as well as for corneocyte desquamation [3]. At the stratum granulosum-corneum junction, the keratinocytes extrude those lipid precursors from the intracellular space into the intercellular lipid space via exocytosis. During this stage, the cell dehydrates and loosens its nucleus, followed by cornification into a corneocyte and SC formation followed by desquamation [21]. Corneocytes adopt a characteristic flattened, dead cell dimension and are oriented parallel to the skin surface that interacts closely with the surrounding grid [11, 22]. Those corneocytes are the building blocks of the skin barrier and are formed by a unique process of programmed cell death associated with the expression of specific genes that control multiple steps of this pyroptosis program. The SC, which is typically 10-20 µm thick, comprises 10-30 layers of corneocytes, which are filled with water and microfibrillar keratin filaments [7]. With a width of 30 µm and height of 1 µm, a corneocyte is embedded in a complex continuous, multilamellar, lipid-enriched matrix. These lipids are enzymatically processed from their precursor form to their final state. Free fatty acids, cholesterol, and ceramides are the main lipid classes inside the epidermis, whereas ceramides account for about 50 % of the SC lipids [12, 13]. The ceramide family comprises at least seven subfractions, they are important not only due to their predominance but also their long-chain and amphiphilic structures and are crucial for the formation and maintenance of the cutaneous permeability barrier of the skin [23]. Ceramides are important as bioactive metabolites involved in cellular signaling, proliferation, differentiation, and apoptosis in the human epidermis and they are essential components of the plasma membrane [24]. Additionally, the SC contains other lipids such as triglycerides, wax esters, and squalenes [21]. This intercellular lipid space constitutes about 15 % of the SC dry weight with about 20 % water content [3, 25]. The SC provides a unique lipid matrix structural basis, which leads to an extraordinary low water and antimicrobial permeability. The corneocytes provide mechanical UV protection [3]. They build a cornified envelope together with a densely crosslinked layer of late differentiation marker proteins such as filaggrin, loricrin, and involucrin and are, along with γ-transglutaminase, responsible for the cornification process [13]. During cornification, the intracellular organelles within the epidermis are replaced by a compact proteinaceous cytoskeleton. This is followed by the formation of the cornified envelope due to cross-linking of proteins at the cell periphery and the linkage of the corneocytes into a functional biologically dead structure [26]. The viable epidermis, which is 50-100 µm thick, is a non-vascular tissue with sensory nerve endings and acts as the first line of immune defense [12, 27]. In addition to the keratinocytes, this epithelial tissue contains other metabolically active cells such as Merkel cells, melanocytes, and Langerhans cells [28]. Merkel cells are approximately 10 µm in diameter and are located in the basal layer, more or less frequent depended on the body region, and differentiate from the ectoderm [29]. Controversially discussed, they are associated with sensory nerve ending function and are assumed to play a role in response to light and touch stimuli [30, 31]. Melanocytes, which are exclusively localized to the basal layer, account for about 10 % of the basal layer cells and derive from the neural crest ectoderm [32]. They share many characteristics with dendritic cells and their dendrites interact with other layers of the epidermis [27]. Their main purpose is the production of three melanin types: eumelanin, pheomelanin, and neuromelanin. Melanin is transported to the nearby keratinocytes to induce pigmentation, which serves as protection against UV radiation [31]. Besides this, melanocytes play a role in the immune system and immune response. However, the primary skin immune response is linked to Langerhans cells [7]. They are present in all viable epidermis layers, but most prominently in the stratum spinosum. Langerhans cells share similarities with both dendritic cells and tissue macrophages, are specialized in antigen presentation and are involved in immune homeostasis and the uptake of apoptotic bodies [31]. Overall, the epidermis is important for a reservoir function and acts as a main barrier against pathogens of the skin. The epidermal differentiation process is linked and regulated inter alia by the concentration of extracellular Calcium and by its acidic cornified envelope, which regulates the homeostasis of the epidermal permeability barrier [3]. The acidic environment (pH 4-6) protects the body by maintaining the natural skin microflora and prohibits the growth of certain pathogenic microorganisms [12]. Many skin enzymes, such as β-glucocerebrosidase and sphingomyelinase, are linked to the SC lipid homeostasis and function optimally under acidic conditions. [33, 34]

Dermis

The dermis is located under the epidermis, has a thickness of 1-2 mm, and is divided into the upper papillary layer and lower denser reticular layer [7]. Those layers are derived from the dermatome region of the mesoderm [35]. The dermis consists of extracellular matrix, which is made up of intertwined elastic fibers and collagen proteins that comprise the bulk of the skin and yields an elastic and tight tissue [12]. There are different collagen types inside the skin, and type I and II account for approximately 75 % of the dermis dry weight [2]. Besides the extracellular matrix, this connective tissue contains sensory nerve endings, blood vessels, and lymphatic vessels [12, 36]. Different cell types including immune cell systems are embedded in the dermis. The most prominent cell type in the dermis is the fibroblast, which accounts for the synthesis of the extracellular matrix and collagen and is derived from the paraxial mesoderm [13, 37]. Dendritic cells and their subsets are involved in regulation and initiation of many immunologic responses and play a role in dermal inflammation, wound healing, skin sensitization, tissue remodeling, and activate the humoral immune response [7, 36]. The dermis also includes hair follicles, sweat glands, and sebum glands, which increases the surface area depending on the body region [38]. They invaginate the skin´s epidermis and dermis and open directly into the skin surface and to the environment [39]. Penetration via these pilosebaceous units is known as the appendage route, and they have been shown to have a complex, dynamical structure and contribute to the regulation of biochemical, immunological, and metabolic activities [7].

Subcutis

The innermost layer of the skin is the subcutaneous layer or hypodermis, which is derived from the dermatome region of the mesoderm [35]. This layer encloses fibroblasts, adipose cells, and macrophages. It is composed of adipose tissue, fascia, as well as larger lymphatic- and blood vessels [12]. The adipose tissue, made up of adipocytes or fat cells is mainly used for fat storage and acts as an insulator to regulate body temperature [40]. It is a loose connective tissue and forms in entirety the panniculus adiposus. The fascia surround and separate muscles, bones, and joint capsules and are an assembly of fibrous, tough connective tissues [7]. The delivery of active ingredients inside the hypodermis and therefore to the muscle and fat tissue is a complex process; thus, direct penetration and systemic blood circulation supply contributes to the muscular deposition [41]. Drug lipophilicity, molecular size, and protein binding ability influence the hypodermis portioning as well as the dermal blood flow in local tissues [42].

4.2 Drug delivery via the skin

Due to the biological and metabolic activity of the skin, an understanding of how substances get into and through each skin layer is essential for a target-site activity. Most APIs penetrate by passive diffusion across the skin, whereas active transport plays a limited role [38]. There are multiple biotransformations such as oxidation, reduction, hydrolysis, and conjugation that occur inside the skin. These processes have minor influences on penetration; for example, steroids (testosterone and estradiol) are converted to an active form while penetrating the skin [43]. Therefore, the bioavailability of topically applied substances might be influenced by biotransformation, enzyme activity, or transport mechanisms [13].

As described by the Organization for Economic Co-operation and Development (OECD), dermal absorption is the global term that describes the transport of molecules from the surface of the skin in the systemic blood circulation divided into permeation, penetration, and resorption [44]. Permeation is the diffusion movement through pores or the entry of a substance into a particular layer such as the entrance into the SC. However, penetration is a process of translocating into deeper layers and going from one layer into a second layer that is functionally and structurally different from the first one. Resorption is the uptake of molecules in the systemic circulation and therefore systemic absorption, which is mostly referred to according by Environmental Health Criteria (EHC) 235 from the World health organization (WHO) as transdermal penetration [45]. The rate and efficiency of API absorption and penetration into the skin is linked to multiple aspects: the ability to pass through the SC, the thickness of the SC, pathological changes in the skin, the number and size of hair follicles, age, and blood circulation [21]. The anatomical location of the penetration site has a strong contribution to its permeability. Follicular dense regions like the forehead, ears, and cheeks show for certain substances an increased penetration rate compared to the forearm [21]. Despite a thicker SC, the penetration rate at the palms is similar to that at the forearm [21]. Furthermore, when considering the permeability of diseased skin, the influence of the altered barrier function needs to be given special consideration [46]. In general, there are three main potential routes for drug delivery into or across the skin: (1) The intra- or transcellular route through the corneo- and keratinocytes. Via this route, drugs pass through both the lipid matrix and cells. (2) The intercellular pathway, where the drug needs to pass through the lipid matrix and the intercellular space between the cells of the skin [47]. (3) Through the transappendageal- or follicle route, the drug penetrates through the hair shaft, sebaceous glands, or sweat glands to get into the skin [48]. The intercellular route is the most common way for drug penetration across the whole skin and involves several steps: (a) dissolution and release from the formulation, (b) partitioning into the SC, (c) diffusion across the SC and the lipid domains, (d) partitioning from the SC into the more aqueous viable epidermis part, (e) diffusion through the viable epidermis over the tight junctions into the dermis part, and (f) drug absorption and uptake in the local target tissue site and vessels, which leads to systemic circulation [49].

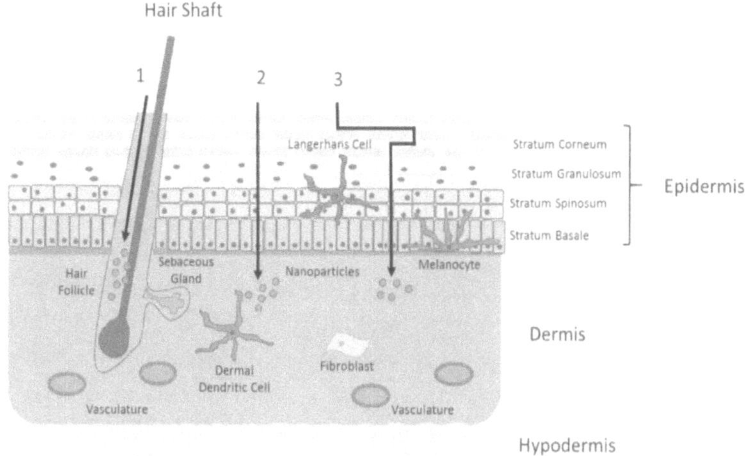

Figure 2: The structure of skin illustrating different penetration pathways. The skin is organized into multiple layers, which contain different cell types. Skin penetration occurs through the follicular (1), intracellular (2), or intercellular (3) pathway. [50]

Penetration inside the skin

The pathway for topical penetration is influenced by potential absorption from the skin surface into the lipophilic SC followed by a subsequent permeation through the more aqueous, non-vascular viable epidermis into the perfused, aqueous dermis [38]. The API in a formulation aims to achieve equilibrium by redistributing itself into the skin. This partitioning of the molecule to redistribute the concentration gradient, formed when the API is applied to the skin, combined with the affinity of the penetrating API to lipophilic or hydrophilic environments is the driving force for penetration into skin [38]. The API diffuses within the formulation to the formulation-skin interface and partitions into the skin. There, it diffuses through the lipophilic SC, thereafter partitioning and diffusing into and through the hydrophilic viable part of the epidermis. From the epidermis it partitions into and diffuses through the dermis before partitioning into the systemic circulation [51]. Lipophilic molecules cross the SC, which is described as the primary penetration barrier and will penetrate slower through the hydrophilic viable epidermis and dermis. The passive penetration of active ingredients inside the skin can be grouped into (1) molecular diffusion, (2) partitioning, (3) metabolism and phase change, and (4) binding and absorption [52]. This passive penetration process has been described as Fick's first law of diffusion in 1855. Fick´s law describes the passive diffusion movement of unbound molecules, towards equilibrium in response to a concentration gradient [53].

$$(1)$$

$$(2)$$

$$J = K_p \times \Delta C$$

$$K_p = \frac{P \times D}{h}$$

The molecule penetration rate is expressed as the flux (J), under steady-state conditions, with the concentration gradient across the membrane (ΔC), and the permeability coefficient (K_p). K_p is expressed by the partition coefficient (P), the diffusion coefficient (D), and the diffusion path length (h).

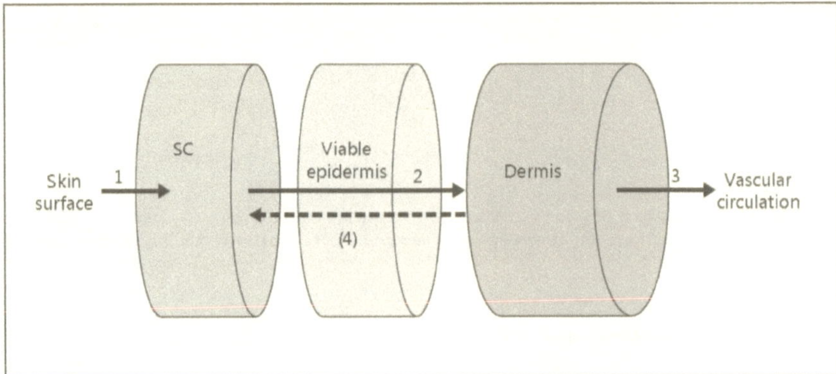

Figure 3: Model of the absorption across the skin barrier. 1 = Penetration phase: passive diffusion into the lipophilic SC; 2 = permeation phase: the transport through the more aqueous, avascular, viable epidermis to the highly perfused dermis; 3 = resorption phase into the microcirculation and further to the systemic circulation or deeper into the tissue (regional penetration); (4) = affinity of the penetrant for the SC or the dermis (reservoir formation). [38]

Epidermis

The main barrier function of the skin is linked to the epidermal layer. Generally, small hydrophilic molecules overcome this barrier via the transcellular route and small lipophilic molecules via the intercellular route [12]. Compared to a physiological pH of 7.4, which is present in the blood stream, the pH of the skin varies within a range of pH 4.5 to 5.5 and is therefore slightly acidic and regulated via endogenous factors, e.g. skin moisture, sebum, nutrients, and age [54]. This acidic environment is crucial for the control of the SC integrity and cohesion as well as for an intact antimicrobial and epidermal permeability barrier [3]. The composition and homeostasis of the skin is crucial for its barrier function and is therefore essential for the penetration ability of APIs [55].

SC

The SC and its heterogeneity ensures the barrier function and acts as the main and primary defensive line for exogenous substances [56]. Therefore, the composition of the SC lipid envelope contributes to the maintenance of the skin barrier function [12]. The lamellar structure of the intercorneocyte space leads with its lipid bilayer and complex non-polar lipids to an intact barrier function. The most abundant SC lipids are ceramides, which are important for water transport and barrier function, which leads to effective epidermal barrier function [57-59]. The SC provides a reservoir function, which is directly linked with its barrier ability over time. By increasing the thickness of the SC, the reservoir function decreases with increasing barrier function [60]. The reservoir capacity of the SC is limited and its saturation is dependent on donor, application site, substance, and formulation [21]. Also, its interaction and affinity to certain drugs provides an accumulation of the active ingredient in the SC, which leads to local action of the drug [21]. It is known that lipophilic drugs like corticosteroids, with a characteristic log P of 1-4, have good skin permeability properties especially across the lipophilic SC environment [28, 61]. Therefore, the SC polarity or lipophilicity is assumed to be 0.8 [51]. Depending on the species, the thickness of the SC and lipid composition varies, with pigs having the thickest and mice the thinnest SC among mammalians. The penetration abilities of APIs correlate to a certain extent to the SC thickness, since several structural differences influence this complicated process [21].

Viable epidermis

The target-site penetration inside the viable epidermis depends on the permeability of the drug across the SC and the impermeability into the dermis. After overcoming the first line of defense, the SC, or the aqueous epidermis, might be the tissue of interest for target-site penetration or needs to be passed through to reach the dermis. Lipophilic compounds that overcome the SC barrier, often demonstrate reduced penetration rates if the more hydrophilic epidermis is reached [38]. Between the viable epidermis and the dermis is the basement membrane located, which builds with the presence of tight junctions a significant permeability barrier that affects the transport of drugs. The barrier function of the basement membrane is mainly impermeable for negatively charged macromolecules and large substances. On the other hand, tight junctions form a permeability barrier for small, intermediate, and large size molecules as well as ions [55, 63]. The stratum granulosum provides with the junction proteins claudin, occluding, and tricellulin a physical barrier for substances [7]. In addition, the lamellar body organelles are considered to be an essential part of the stratum granulosum barrier function with a closely linked antimicrobial function of antimicrobial peptides like cathelicidine [3]. The lamellar bodies are important for the formation, maintenance, and repair of epidermal barrier function. In addition to their contribution to the epidermal permeability barrier they influence the defensive function exerted by the SC. The drug binding capacity, sequestration, metabolism, and active transport in the layers of the viable epidermis harms the availability of active ingredients inside the dermis [64]. Owing to its role as the first line of immune defense, there is a need for direct delivery of the drugs to certain skin layers. The lower epidermis contains melanocytes and Langerhans cells and is therefore a local target-site and specific penetration to this area by avoiding transport through the dermis is desirable [27]. [3, 7]

Dermis

The metabolic activity and capacity of the epidermis exceeds that of the dermis [13]. Divided into the upper papillary and the lower reticular layer, the upper dermis exhibits thinner and looser packing of collagen fibers. The papillary part interdigitates with the basal layer and builds a dermo-epidermal junction [12]. After the API passes the aqueous epidermis and enters the aqueous dermis, it penetrates the tissue and moves through the layers via passive diffusion. In the dermis, active ingredients may enter the blood stream or the lymphatic circulation, which leads to systemic exposure. The microvasculature and lymphatic system is mainly present at the top 1-2 mm of the skin, due to which drugs are cleared from their target-site via systemic transport [65, 66]. The transfollicle route is important to accumulate drugs in the dermis, which enables drug targeting with a sustained effect [67]. Regional accumulation inside the dermis is indispensable for the prevention and treatment of local skin diseases [7].

Appendages

The surface area containing hair follicles is highly dependent on the body region and extends the skin area as well as the penetration area from a maximum of 13.7 % at the forehead to a minimum of 1 % at the forearm [68]. This variance is caused due to different volumes and the different amount of hair follicles at different body sites. The hair shaft increases the epithelial surface area and confers a reservoir function as well as an interface for the interaction of drugs. Physico-chemical drug properties (lipophilicity and molecular weight), vehicle, and formulation types are highly important for its penetration ability and also influence the distribution of the drug [69]. The lower part of the infundibulum acts as a main entry point for drugs, since the barrier function of the hair follicle is interrupted [7]. In this upper appendage shaft, there is a dense capillary network, which facilitates drugs to enter the systemic circulation more effectively. These transappendageal pathways have gained increasing interest and have been for a long time overlooked as an efficient penetration route. Increasing evidence has revealed the importance of hair follicle penetration as a fast entry route into the skin and is now being discussed as a shunt route [7]. The transfollicular pathway allows for enhanced penetration into deeper sites, prolonged residence duration, and site-specific targeting [70]. In contrast, other appendage pathways such as the sebaceous and sweat glands have been poorly investigated, due to difficulties with precise local detectability [71]. Exclusive penetration via the hair follicle can be studied and blocked by using nail varnish; Klein et al. demonstrated that nail vanish provides secure follicular closure by not increasing the penetration of the active ingredient [72, 73]. There are active follicles inside the skin, which secret sebum and are local points for hair growth and passive follicles, but their activity remains unknown this far. Only active follicles are relevant for topical skin penetration [74]. Physiological factors such as hair growth cycle and regional variation affect follicular penetration [39, 75]. The sebaceous glands create a neutral and non-polar lipid environment, which offers a potential penetration pathway for lipophilic molecules [39]. Lipophilic formulations and the physico-chemical character of the active ingredient play an essential role in follicular trafficking, and some drugs are known to exclusively penetrate via this route [76]. The physico-chemical properties of hydrophilic molecules (caffeine), electrolytes, and larger molecules favors transfollicular penetration.

Lateral diffusion

Recent research states that lateral diffusion needs to be considered in penetration studies and more passive API movement takes place than previously thought. APIs are spreading on top and within the upper layers of the SC. Predominantly lipophilic molecules diffuse laterally within the SC [77]. This movement is restricted by a topically applied fence around the application site, which leads to decreased spreading of the active ingredient. A recent study showed that lateral penetration of APIs within the viable epidermis and dermis has an impact on the overall penetration within the skin [78]. To better understand this concept, penetration models need to be updated to be able to study lateral penetration.

4.3 Penetration models

The penetration ability and percutaneous absorption of the skin varies largely with age, sex, ethnicity, size, weight, and body region [79]. Penetration models are important to investigate the efficacy and quality of topically applied active substances in a formulation [2]. To study the dermal and transdermal delivery of drugs through the skin *in silico, in vitro, ex vivo* and *in vivo* experimental studies are accomplished [38].

In silico skin models are used to predict the permeability of active compounds. This approach is based on theoretical-mathematical simulations derived from structure-based permeation models. Experimentally validated data correlate with the concentration-depth profiling of SC penetration simulations. However, the use of molecular dynamic simulations is limited and provides a first simple assumption of penetration abilities. [80]

In vitro skin models are designed for transdermal drug delivery testing with the utilization of artificial membranes or reconstructed human skin. Artificial membranes are synthetic skin alternatives containing silicone, multiple layer polyether sulfone, or synthesized copolymer membranes. The Strat-M® membrane is a non-animal based, synthetic artificial membrane, which is built up in multiple layers of polyester sulfone, coated with lipids, specially designed to mimic the skin structure. Artificial membranes have the potential to be screening models, by providing trends and correlations for penetration enhancement ability. They provide a beneficial low intermembrane variation, which leads to better trend estimation of API and formulation penetration abilities. [81]

Apart from skin-mimicking artificial membranes, some *in vitro* models are based on reconstructed skin, using human epidermis and dermis cells. There are multiple models such as Graftskin™, SkinEthic™, HRE, and Episkin that have been established for penetration testing [82, 83]. Those skin models are limited as surrogate models due to their high penetration ability [84, 85]. Both *in vitro* skin model systems have a lower barrier function, which leads to higher penetration rates [86]. Nevertheless, artificial membrane *in vitro* assays are a valuable tool to predict penetration.

In vivo human penetration studies represent the most accurate method to predict efficient penetration abilities with *in vivo* relevance [87]. There are multiple options on *in vivo* human penetration studies. For example, a biopsy removal as an indirect *in vivo* method provides valid information about a substance's depth profile inside the skin. After topical application, the SC is fully removed via tape stripping [88] or cyanoacrylate stripping [89], or the full epidermis is removed via heat-separation [90] for depth profiling. This invasive method is limited due to ethical reasons. *In vivo* confocal laser scanning microscopy provides valuable information about three-dimensional skin structure and the localization of an active ingredient inside the skin [91]. This technique is limited to fluorescence substances which are semi-quantitatively coupled to the active ingredient. Another technique for depth profiling is *in vivo* Raman imaging [92]. APIs are localized and significant skin structures are detected via semi-quantitative analyses. The minimally invasive microdialysis *in vivo* setup is used to identify the penetration of active ingredients into the skin and does not show any swelling or onset of inflammation due to insertion of the membrane [93, 94]. Due to ethical reasons and limited access to *in vivo* human skin, penetration experiments with different alternative models have been proposed. As an alternative skin model, the skin from pigs, rats, hairless mice, rabbits and snakes has been used [95]. *In vivo* studies have established a penetration rate trend for percutaneous substances: rabbit > rat > porcine skin > human [96]. To compare the absorption rate of different skin species, the penetration rate needs to be considered. Due to ethical reasons and the derived animal ban restrictions in the European Union for cosmetic ingredients and products, animal testing has been prohibited since 2013 [97]. The use of animal waste products from food production is a permissible alternative method. An alternative to *in vivo* human studies is the use of *in vitro* and *ex vivo* skin for percutaneous penetration testing, which is accepted by the Scientific Committee on Consumer Safety (SCCS) [98].

Ex vivo human skin, which is obtained from plastic surgery, is regarded as most representative to *in vivo* human skin. Excised human skin is limited and difficult to access [21, 99]. The OECD guidelines recommends porcine ear skin as a suitable model for percutaneous penetration studies [44]. The composition of the SC and epidermis as well as the lipids and hair follicle density inside the skin is among other factors important to compare penetration rates. The follicle density is dependent on the species and most animals exhibit a higher density than humans [68]. Considering human mimicking models, porcine ear skin has similar dimensions of hair follicles but with a larger diameter [100]. In addition to the follicle density, the morphology and penetration abilities of porcine ear skin are comparable to those established *in vivo* human studies for topically applied substances [101, 102]. The SC shows a uniform arrangement of corneocytes with a similar number of cell layers [21]. Compared to the whole porcine body with a thickness of 17-28 µm for the SC and 60-85 µm for the viable epidermis, the SC of the porcine ear is with about 19 µm rather thin [103, 104] and with about 24-38 % of total epidermis thickness comparable to the 16.8 µm SC of humans [95]. In general, the absolute thickness of the epidermis and dermis of the porcine ear skin corresponds to that of the human skin (Table 1) [103].

Table 1: Comparison of human and porcine skin layers with average thickness [105].

	SC	Epidermis	Dermis	Reference
Human	16.8 µm	70 µm	1.20 mm	Qvist *et al.* (2000) [106]
				Bronaugh et al. (1982) [95]
Pig	19 µm	72 µm	1.86 mm	Jacobi *et al.* (2007) [104]
	(17-28)	(60-85)	(1.44-2.25)	

For penetration testing, *ex vivo* human and porcine skin models are the specimens of choice [107]. Porcine ear skin is reported as more suitable than excised human *ex vivo* skin due to the presence of cartilage inside the ear, which leads to less contraction and a more stable tissue [108]. For *ex vivo* percutaneous penetration studies, separation of the skin into the SC, epidermis, and dermis, post experiment provides a suitable penetration profile of the concentration of the compound per layer. The passive diffusion of APIs is experimentally studied using *ex vivo* and *in vitro* percutaneous penetration diffusion cell models. Two types are commonly used, the static or flow-through diffusion cell [38]. The flow-through penetration system provides continuous replacement of the receptor fluid (RF) in contrast to the static diffusion cell system. Using surrogate skin systems to investigate penetration, the Franz Diffusion Cell (FDC) has been applied since 1975, and is the most commonly used static diffusion cell setup [109]. This method with its simple design is used for *in vitro* and *ex vivo* penetration studies. The cell system consists of a receptor chamber filled with a RF in which the compound is released after penetrating through the skin. The RF is constantly stirred and the receptor chamber is temperature controlled at 32°C for cosmetic studies and 37°C for dermatological or pharmaceutical penetration studies. For *in vitro* penetration studies artificial membranes, and for *ex vivo* penetration studies, skin is mounted on the receptor chamber. When using skin, either full-thickness or split-skin is used as the surrogate system. Most percutaneous penetration studies use 200-500 µm dermatomed split-skin, which exhibits only the upper part of the dermis, making the difference in the dermis thickness less important for interpretation of penetration ability [110]. The donor chamber is placed on top of the skin to ensure correct topical application of either an infinite dose (>100µL/cm²) or finite dose (10µL/cm²) [105, 111]. This setup allows the evaluation of concentration kinetics as well as determination of the absolute amount of an active ingredient within different skin layers. Cumulative sampling allows for analysis of the kinetic progression of penetration, and by individual sampling, the total amount of penetrated API is determined. Penetration kinetics allow to predict the bioavailability, which is important to ensure the efficacy of the compound exposed to the living cell entity targeted by the API. In order to determine the concentration in the respective skin depth, the skin is separated into different layers and the active amount is analyzed. The major disadvantage of the FDC is the invasive preparation of the skin samples, which possibly results in trauma and thus does not adequately mimic healthy skin conditions [112].

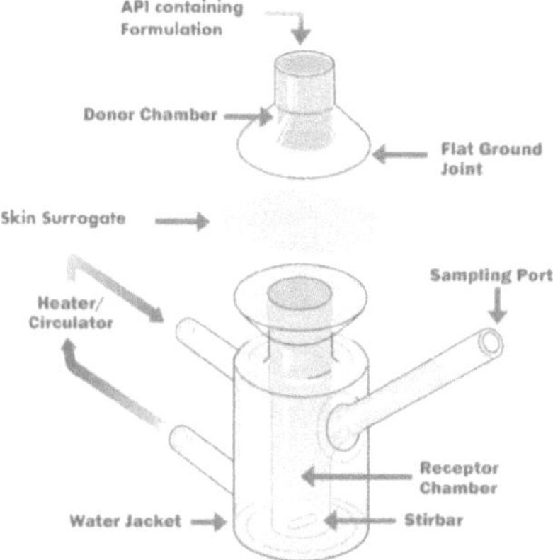

Figure 4: Schematic illustration of a the FDC. Figure modified by M. Lubda. [113]

Flow-through diffusion cells for skin research have gained popularity within the field of neuroscience. The microdialysis (MD) system, which is the most used flow-through system, is used in dermatology research. The continuously replaced RF mimics the *in vivo* circulation of blood inside the skin. A formulation containing the active ingredient is topically applied on the skin, and then penetrates and diffuses inside the different tissue layers. The application site is defined with a fence to ensure a spreading within the desired area. A semi-permeable membrane, which is placed underneath the skin is flushed with a perfusate, mimicking a small blood vessel in the skin [114]. The active ingredient, which penetrates the skin, is collected inside the membrane and flushed by the perfusate, enabling sampling. The flow rate is an important factor limiting the penetration and affecting the relative recovery, which is an indicator of the performance of the MD system. The higher the flow rate, the higher the recovery rate, but the less active ingredient is found in the perfusate; a flow rate between 1 and 5 µL/min is commonly used to ensure a steady state diffusion into the membrane [115]. The MD setup allows to detect small molecules, like APIs as well as endogenous molecules such as cytokines [116, 117]. The membranes are defined by their molecular weight cut-off, which is an important factor. Therefore, the molecular weight cut-off should be 10-times larger than those of the collected molecules, to achieve a high recovery of the active ingredient [116, 117]. While performing the MD experiments, the localization of the membrane inside the skin is important. The membrane is placed in a defined depth, which enables the determination of the penetration depth for each respective substance. In recent research, the depth has been controlled via ultrasound after the experiment has been carried out [118]. Ideally, depth validation is performed prior to the experiment. Using ultrasound, the skin is treated with gel to receive a detailed image, but the gel has an impact on the penetration. A not yet described alternative method, is the depth analysis determined via computed tomography. This non-invasive method allows to determine the depth of the membrane before the experiment. MD has several benefits, including its minimally invasive experimental setup and continuous monitoring of the pharmacokinetic and pharmacodynamic studies of *in vivo*, *ex vivo*, and *in vitro* experimental setups [119].

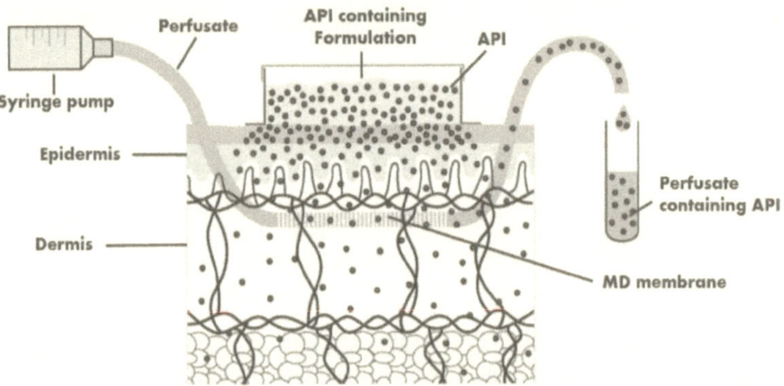

Figure 5: Schematic experimental setup of the microdialysis. Figure modified by M. Lubda. [120]

The perfusate which is flushed through the MD membrane and the RF of the FDC needs to dissolve the active ingredient without degrading it. This fluid is a physiological solution that mimics the blood or the skin environment [121]. In most percutaneous penetration studies, phosphate buffered saline (PBS) or Ringer Solution is used for hydrophilic molecules as the physiological fluid. For molecules with a lipophilic character, the perfusate needs to be optimized to ensure the solubility of the active ingredient. The recovery of APIs is increased for lipophilic molecules, by adding solubility enhancers like albumin, cyclodextrines, or the surfactant Brij [105, 122, 123].

The barrier function of skin is commonly linked to transepidermal water loss (TEWL), which describes the loss of water in gram per surface area over time and provides information about the integrity of the skin. To ensure the integrity of the skin, TEWL, infrared spectroscopy, and impedance spectroscopy are used to distinguish the barrier function of the skin [124-126]. The determination of the TEWL occurs prior to the penetration experiment and is the commonly used method to analyze barrier-disruption of the skin [3]. The post-experimental quantification of the active concentration is performed using HPLC.

4.4 Influences on the penetration of the API

The kinetics of percutaneous penetration studies and the ability of an API to reach its target-site is influenced by multiple factors. The penetration ability is dependent on the physico-chemical properties of the API itself. Besides the characteristics of the API, the formulation or vehicle components, which surround the active ingredient, influences the penetration ability. External factors and application of the API containing formulation influences the penetration depth and rate.

API

The passive diffusion of most APIs follows Fick´s law, and therefore, the physico-chemical properties of APIs are important and influence the permeability coefficient, which modulates its penetration ability [38]. In particular, the molecular size analog to Stokes surface, expressed as molecular volume and molecular weight and the lipophilicity, which is expressed with the log P value, influence the penetration [127].

As a general rule, molecules with a molecular weight smaller than 500 Da penetrate the skin passively [62]. Molecules with a higher molecular weight have the potential to passively or actively penetrate the skin, but to increase the penetration rate, an electrical or mechanical enhancement is recommended [128]. With increasing molecular weight, the penetration rate decreases, defined by the diffusion coefficient. The lipophilicity is defined by the octanol-water partition coefficient ($K_{o/w}$), describing the concentrations of the dissolved molecule in the organic octanol and aqueous water phase [129]. This coefficient is expressed as log P and positive values indicate lipophilicity and negative values hydrophilicity. Therefore, a log P of 1-3 permeates skin most readily [12] and APIs need a lipophilic characteristic to enter the SC and a hydrophilic characteristic to pass through the viable epidermis [49]. The log P describes the solubility characteristics of an API. To ensure penetration, the API needs to be dissolved in the formulation. The solubility of an API is linked to its melting point; as a rule of thumb, active ingredients with a melting point of less than 200°C are said to penetrate the skin [130]. A high solubility leads to a high concentration gradient, which increases the diffusion force across the skin. The penetration rate of an active ingredient is optimized by a low molecular weight, a low melting point, and moderate lipophilicity. Besides this, the main characteristics of an API such as the vapor pressure, the ionization with its pH dependency, hydrogen bonding activity, and the affinity for protein binding affects the percutaneous penetration [12, 38].

To study the main effects of the physico-chemical influences on skin penetration of APIs, the molecular size and lipophilicity of compounds are tested. Caffeine is recommended by the OECD guidelines as a model compound for *in vivo, ex vivo,* and *in vitro* penetration studies [44]. The methyl xanthine alkaloid caffeine is hydrophilic in nature, moderately soluble in water and well known for its penetration behaviour [131]. Furthermore, the physico-chemical properties of the active ingredient LIP1 are of interest but its penetration abilities are unknown. LIP1 is a lipophilic molecule with a low molecular weight and low solubility in water. Caffeine and LIP1 have a similar molecular weight with a difference in the lipophilic characteristics, which makes them ideal for comparative penetration studies (Table 2). The penetration is influenced by the physico-chemical characteristics of the molecule itself as well as the vehicle and external factors.

Table 2: Physico-chemical properties of caffeine and LIP1. [131]

	Caffeine	LIP1
Chemical structure		
Molecular weight [g/mol]	194.19	194.19
Log P	-0.1 (exp.)	0.6 (exp.)
Melting point [°C]	236 (exp.)	179 (exp.)
Solubility in H_2O [25°C]	21.6 mg/mL (exp.)	0.75 mg/mL (exp.)
Solubility in PGOA [25°C]	7.2 mg/mL (exp.)	52.0 mg/mL (exp.)
CAS number	58-08-2	65609-28-1

Formulation

To ensure target-site specific penetration of APIs, the formulation, which is described as the vehicle of the API, is important. The formulation allows to optimize the delivery of active ingredients into the skin and needs to be designed individually for each API [51]. Most APIs are formulated into a standard formulation, almost exclusively based on their physico-chemical properties and solubility characteristics. Due to complex formulation processes, formulation development is initiated after the efficacy testing. Many different compounds, such small hydrophilic and lipophilic molecules, surfactants, lipids, and solvents are part of a formulation. They are divided into different classes and most of them have a penetration enhancing effect and improve the penetration of the API. Penetration enhancers increase the penetration rate of APIs in different ways but should ideally not irreversibly change the skin morphology. Therefore, the following hydrophilic and lipophilic penetration enhancements are described [132]. Hydrophilic penetration enhancers interact with the hydrophilic groups of the phospholipid bilayer, which leads to reorganization of the lipid chain, due to unbound water. Lipophilic penetration enhancers loosen the parallel structured lipid chains. Some solvents increase the solubility of the API in the aqueous phase or oily phase, which leads to a higher concentration gradient. [8, 51, 133]

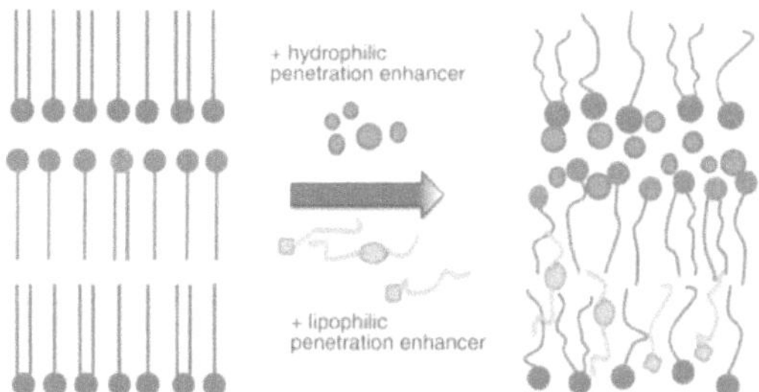

Figure 6: Lipid bilayer of the SC and the enhancing effect on the penetration of hydrophilic and lipophilic penetration enhancers. [132]

Penetration Enhancers

Emollients:

Most emollients are added to a formulation as a refatting agent, they are substances that soften the skin and slow down the evaporation. Beeswax, almond oil, paraffin, and glycerol are commonly used emollients. They have a broad penetration enhancing effect, and polar emollients are particularly known to promote API penetration. [134]

Surfactants:

Surfactants are detergents present in multiple skin products. They alter the TEWL and the transition temperature of SC lipids and disturb the SC diffraction patterns [135]. Surfactants such as sodium dodecyl sulfate are known for their penetration enhancing function, which elicits an irreversible effect. They affect both the intra- and intercellular pathway and lead to SC swelling [136].

Azone[*]:

Laurocapram or Azone[*] is a hybrid of pyrrolidone and decylmethylsulfoxide, which are potent penetration enhancers. This lipophilic molecule is especially designed to achieve a skin penetration enhancing effect. Azones[*] interrupt the lipid packing by partitioning into the lipid bilayer and increasing the skin permeability for hydrophilic and hydrophobic drugs [137]. Combined with propylene glycol, Azones[*] are known to enhance the penetration flux of actives across the skin [138].

Humectants:

Those osmolytes are small, polar compounds like urea and glycerol, are naturally present in the SC and are known to moisturize the skin [139, 140]. They cause increased mobility in the ceramide headgroups, which leads to a penetration enhancing effect, paired with hydrophilicity [141]. Humectants partition into the aqueous regions between the lipid lamellas and therefore loosen the lipid packing, as well as inside the corneocyte [8].

Glycols:

Glycols are commonly part of the aqueous phase of a formulation and are known to have a penetration enhancing function [142]. In addition to their penetration enhancing effect on the skin, they increase the exposure of the API to the solvent, thereby increasing its solubility. Propylene glycol (PG), butylene glycol, ethoxydiglycol, and glycerol are commonly used glycols and are present in different formulations. PG as a co-solvent is the most commonly used glycol and is combined with other penetration enhancers like fatty acids for percutaneous penetration study formulations [143]. PG increases the API solubility and its penetration enhancing function is based on interactions with the polar headgroups or the hydrocarbon chains of the SC bilayers [143].

Fatty acids:

There are two different types of fatty acids: saturated and unsaturated fatty acids. They are the main components of the SC lipids with a long acyl chain and are known to increase the penetration into the skin [144]. The unsaturated form of the acyl chain promotes penetration across the skin more than the saturated form when comparing chains of the same length. Fatty acids affect the mobility of the SC lipids, especially of cholesterol and ceramide headgroups with high water content. They are used to increase the penetration rate of hydrophilic and hydrophobic APIs [8]. Fatty acids such as oleic acid (OA) are known for their penetration enhancing effect; they lower the barrier function while interacting with the polar and non-polar groups of the SC lipids [143].

External factors

External factors, such as occlusion, temperature, mechanical manipulations, and physical forces have an impact on the API penetration into the skin.

Occlusion:

Occlusion of the skin is achieved by covering the skin with a vapor-impermeable membrane or by infinite dosage [87]. This leads to differences in the hydration, barrier permeability, differences in epidermal lipid synthesis and differences in molecular and cellular processes [3]. An infinite amount of formulation leads to a steep concentration gradient and hydration of the SC due to an occlusive effect: both increase the penetration of active ingredients. Engorged corneocytes, increased temperature, and increased hydrostatic pressure from large application volumes causes occlusion which in turn increases the ability of an API to penetrate the skin [145]. Due to a mobile disorder and hydrated conditions the lipids and protein components of the SC change their fluidity [146].

Temperature:

Increasing skin temperature due to either external or internal influence increases the permeability of skin due to occlusion [147]. This can be due to vasodilation, which leads to higher blood circulation and higher evacuation of the active ingredient. At room temperature, the main parts of the SC lipids and proteins are solid; with an increasing temperature they become more fluid, which leads to increased permeability [8]. Increasing the temperature from 25°C to 40°C changes the skin permeability and significantly increases the penetration flux by more than 3-fold [148].

Physical forces:

Electrical voltage achieved via iontophoresis or electroporation, ultrasound by sonophoresis, and microneedles are used to overcome the barrier function and increase skin penetration [149]. Microneedles, with their needle-like microstructures, increase site-specific drug delivery by controlling the length of the needles without invasive stress [150]. These techniques are designed to enhance skin penetration and disrupt the skin without long-term compromise of its barrier function.

5 Materials & Methods

5.1 Materials

5.1.1 Reagents

Caffeine, LIP1, Milli-Q® (H$_2$O), acetonitrile, neutral buffered 4 % Formaldehyde solution, paraffin pastilles, Neo-Clear® and eosin Y-solution 0.5 % alcoholic were purchased from Merck KGaA (Darmstadt, Germany). Propylene glycol (PG), Oleic acid (OA), and Hematoxylin Meyer solution were purchased from Sigma Aldrich by Merck KGaA (St. Louis, USA). PBS with magnesium and calcium (PBS++) was purchased from Biowest (Nuaillé, France). Window color was obtained from Marabu GmbH & Co. KG (Tamm, Germany) and cyanoacrylate super glue by UHU GmbH & Co. KG (Bühl, Germany).

5.1.2 Formulations

Caffeine-H$_2$O solution

Caf-H$_2$O and Caf-PBS were prepared with 2 % or 0.7 % caffeine (w/w), respectively.

Caffeine-PG and Caffeine-PGOA solution

Caf-PG and Caf-PGOA with 5 % OA (w/w) were prepared with a concentration of 0.7 %.

LIP1-PG and LIP1-PGOA solution

LIP1-PG and LIP1-PGOA with 5 % OA (w/w) were prepared with a concentration of 0.7 %.

5.1.3 Materials

Linear EP Low Flux Probes were purchased from EP Medical (Copenhagen, Denmark). Guide canula (G18, 0.60 mm × 25 mm, 100 Sterican®), Omnifix-F 3mL syringes and Perifix® catheter connectors (latex free, 19G, 1.05 mm) were purchased from B. Braun (Melsungen, Germany). Univentor 864 Syringe pump by Univentor (Zejtun, Malta). Makro disposable cuvette were purchased from BRAND GmbH + Co. KG (Wertheim, Germany). Eppendorf and PCR tube were purchased from Eppendorf (Hamburg, Germany). Tesafilm® crystal clear 4129 were purchased from Beiersdorf (Hamburg, Germany). Millex® syringe filter 0.22 µm PES and Strat-M® membrane were purchased from Merck KGaA (Darmstadt, Germany). Cotton was purchased ebelin dm-drogerie markt GmbH + Co. KG (Karlsruhe, Germany).

5.2 Methods

5.2.1 Artificial membranes

The Strat-M® membrane is a synthetic, non-animal-based model, which comprises multiple layers of polyestersulfone and is coated with lipids that mimic the skin structure for penetration testing.

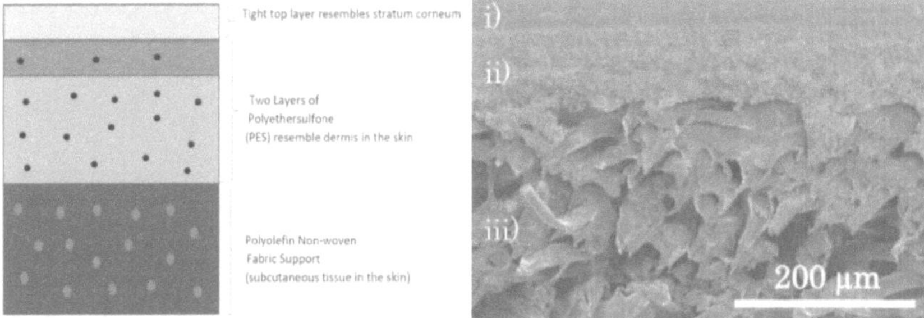

Figure 7: Multilayered structure of the Strat-M® membrane. Scanning electron microscopic image of a cross-section of Strat-M®. First (i), second (ii), and third layer (iii). [81, 151]

5.2.2 Porcine skin

Porcine ear skin was used as an *ex vivo* model for human skin as described in the OECD and SCCS guidelines. Pig ears (German domestic pigs, 6-month-old) obtained from a local slaughterhouse (Brensbach, Germany), freshly slaughtered, were washed and cleaned with rinsing water and then dried using soft tissue and stored at +4°C for a maximum of 72 h. The skin from the back of the ear was dermatomed with an electrical dermatome from Humeca BV (Borne, Netherlands) at a thickness of 300, 500, 700, and 1000 µm, respectively. Six skin punches with a diameter of 26 mm were obtained from each ear.

5.2.3 Human skin

The human skin was obtained from Genoskin (Toulouse, France), stored at -20°C until a maximum of half a year. Due to company safety reasons, fresh *ex vivo* skin could not be used for penetration testing. The skin obtained from the abdomen of female donors was dermatomed with an electrical dermatome at a thickness of 500 µm. Six skin punches with a diameter of 26 mm were obtained from each donor.

5.2.4   Skin thickness

Micrometer screw gauge measurement

To determine the thickness of the dermatomed skin discs for use in the experiments, the thickness was monitored by making four thickness measurements at the outer edge of each disc using an electrical digital micrometer (0-25 mm/0.001 mm) from Vogel Germany GmbH & Co.KG (Kevelaer, Germany).

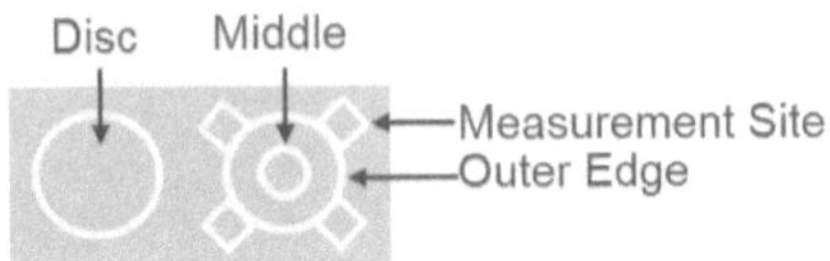

Figure 8: Schematic diagram of dermatomed skin representing each skin disc (Disc) used for the penetration testing and the 9 mm application site (Middle) with four outer edge measurement sites (Measurement Site) to determine the mean thickness of the disc.

To enable this, the skin samples were placed between two cover slips and the thickness was measured using the micrometer considering the first point of physical contact to avoid compression of the tissue.

Microscopic measurement

To determine the thickness of the dermatomed skin with a microscope, small samples next to the outer edge of each stamped disc were obtained and cut into 10-µm thick vertical tissue sectionings using a Leica CM1520 cryo-microtome from Leica Biosystems (Nussloch, Germany) at -30° C. The skin thickness was measured using a bright field microscope (Carl Zeiss, Munich, Germany) and the pictures were analyzed via the ZEN Blue Edition 2.1. software by (Carl Zeiss, Munich, Germany).

5.2.5   TEWL

The TEWL measurement was performed with an open chamber TEWL device CORTEX system from BONDERM GmbH (Biberach, Germany) and each membrane and skin disc were measured five times.

5.2.6   Franz Diffusion Cell (FDC)

Six skin discs/artificial membranes were placed onto the FDC (9 mm diameter, diffusion area 0.636 cm²) from Logan Instruments Corp. (Somerset, USA). The skin was placed dermal side down and the artificial membrane shiny side up onto the receptor chamber. The receptor chamber filled with the physiological buffer PBS was temperature controlled at 32 ± 1°C and constantly stirred. Before placing the donor chambers onto the discs, the TEWL was measured monitoring room temperature and humidity. After an equilibration of 30 min, the formulation was topically applied with 10 µL/cm² (finite) or 786 µL/cm² (infinite). At defined time points, samples were obtained using a Millex® syringe filter unit.

5.2.7 Skin separation

The SC, viable epidermis (E), and dermis (D), were separated for individual skin layer analysis. After each FDC experiment, the donor chambers were washed with 2 mL PBS. The formulation residues on the skin were removed with cotton wipes, applying constant pressure with a roller from Speedball/Brayer (Statesville, USA) and collected in 2 mL tubes; this was termed skin wash (SW).

<u>SC removal:</u>

The SC was removed by five cyanoacrylate superglue tape stripes, which were homogeneously applied with a constant pressure by a roller. The application site was marked on the skin surface. After incubating for 2 min, the tape strips were removed against the grain of hair growth. The demarcated area was punched out, with a diameter of 9 mm and collected in a 2 mL tube; all five stripes were determined as SC. The remaining tapes were collected in a 5 mL tube for lateral diffusion analysis (SC-lat).

<u>E/D removal:</u>

To separate the E and D, the application site was punched out of the skin discs. These were placed D side down on a 60°C heating plate from IKA® (Staufen, Germany) for 90 s and the E was carefully peeled off with forceps. The E and D were weighed and collected in 2 mL tubes each. The remaining skin was collected in a 2 mL tube (Lateral).

The donor chamber wash (DC), skin wash (SW), SC, viable epidermis (E), dermis (D), receptor fluid (RF), lateral diffusion in SC (SC-lat) and skin discs (Lateral) were obtained, respectively. Adding 1 mL PBS and a 5 mm metal ball, the tubes with the skin tissue were shaken in a TisseLyser II from Qiagen (Hilden, Germany) for 10 min at 30 Hz two times and the balls were removed afterwards. Adding to all tubes 2 mL, except for the SC-lat 5 mL, and for the Lateral 1.5 mL, PBS in total, respectively. The samples were incubated for 20 h on a tube roller from Stuart equipment (Staffordshire, UK) at 60 rpm and filtered subsequently.

5.2.8 Hematoxylin-eosin (HE) staining

Preparation of the Paraffin Samples

For the skin separation morphology images, the skin tissue was removed from the cartilage. The skin was fixed in formalin by incubation in buffered formalin (4 %) for 48 h at RT. For paraffin infiltration, the skin was transferred to ascending concentrations of ethanol for 1 h each at RT in the sequence: 70 %, 2x 95 %, 2x 100 %. They were transferred into a xylene solution incubating twice for 1 h at RT. The tissue was incubated in paraffin wax at 56°C over night and after this was repeated 2x for 1 h and the tissue was then embedded in paraffin blocks.

Tissue Sectioning and Staining

The paraffinized skin was cut into 4 µm vertical sections using a CUT6062 microtome from SLEE medical GmbH (Mainz, Germany). The sample HE staining was performed according to the following protocol.

Table 3: HE staining protocol for porcine tissue.

Chemicals	Incubation Time [min]
NeoClear	5
NeoClear	5
100 % EtOH	2
96 % EtOH	2
70 % EtOH	2
Aqua dest.	2
Aqua dest.	2
Haematoxylin	3
H_2O	1
H_2O	3
70 % EtOH	2
Eosin	0.5
96 % EtOH	1
96 % EtOH	2
100 % EtOH	2
NeoClear	5
NeoClear	5

The images were visualized using a brightfield microscope (Carl Zeiss, Munich, Germany) and the pictures were analyzed via the ZEN Blue Edition 2.1. software provided by Carl Zeiss (Munich, Germany).

A	Untreated	B	SC removal

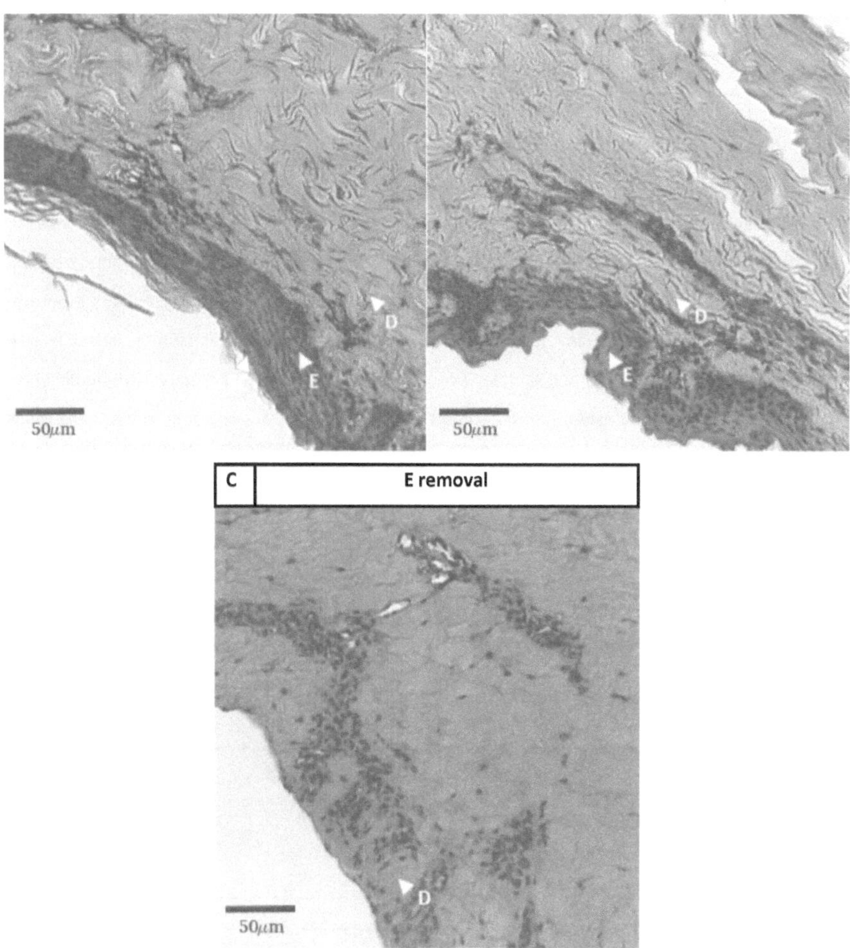

Figure 9: HE stained porcine full-thickness skin after skin layer. A: untreated skin. B: Skin after stratum corneum (SC) removal by cyanoacrylate tape stripping. C: Skin after heat-separation of Epidermis (E) and dermis (D). The SC, E, and D are marked with arrows. 4 µm sections. Scale bar = 50 µm.

5.3 Microdialysis (MD)

5.3.1 Implantation of the membrane in the skin

The porcine ear was pinned dermal side down to a solid surface and the hair was cut manually. The cannulas used as insertion guides were implanted and a 1 cm² application site was defined with window color fence as the site of topical administration with 1 cm between each application site. The membranes were inserted in the skin with a defined depth and a defined length of the membrane accessible to dialysis (insertion length). The membranes were continuously perfused with PBS and an experimental flow rate of 3 µL/min throughout the experiment. The membranes were primed with a flow rate of 25 µL/min for 20 min and for 30 min with the experimental flow rate outside the skin. After insertion into the cannula the membrane was primed with the experimental flow rate for 10 min, the cannula was retracted and primed implanted in the skin for 30 min. The membrane was implanted in the middle of the application site and for lateral penetration studies two membranes were placed additionally at the edge of the application site. The skin was allowed to equilibrate before topical administration of 10 µL/cm² formulation. After the application the perfusate was collected in PCR tubes at the probe outlet and sampled every 30 min for 8 h under tracked relative humidity and room temperature conditions. All experiments were performed by the same operator in the laboratory.

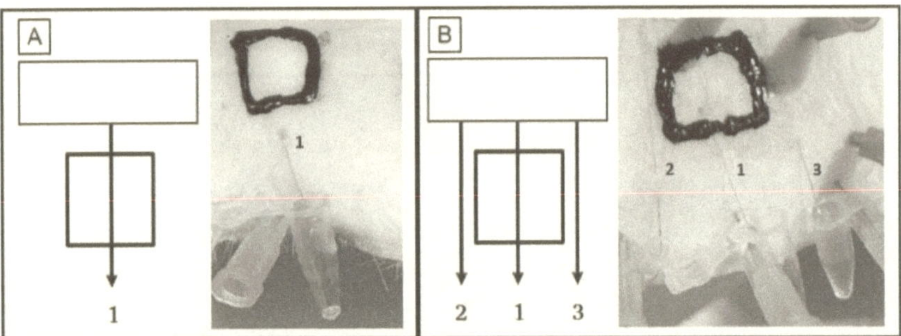

Figure 10: Schematic experimental setup with application site and membrane. A: Experimental setup with membrane in the middle (1) of the application site. B: Experimental lateral setup with membrane in the middle (1) and two membranes (2 & 3) at the edge of the application site.

5.3.2   MD membrane:

The 2 kDa Cellulose cuprammonium (Cuprophan) membrane ($\varnothing$ = 215 µm) with micropores were purchased from EP Medical (Copenhagen, Denmark) were used in MD.

5.3.2.1   Membrane surface structure visualization

To visualize the MD membrane scanning electron microscope (SEM) pictures were taken with a Supr35 Zeiss device by Carl Zeiss (Munich, Germany). The membrane Probes where sputtered with a platinum cover of 10 nm with the Sputtercoater Q150T ES from Quorum Technologies Ltd (Laughton, UK) and the measurement were done under vacuum control and with 30 mA measurements.

Figure 11: SEM images from the 2 kDa MD membrane.

5.3.3   Membrane depth measurement

The depth in the MD experiments in which the membrane was inserted inside the skin was determined via computer tomography (CT), ultrasound, and microscopically.

5.3.3.1   Measurement via CT

The depth of the linear probe was determined with the SkyScan 1176 from Bruker Corporation (Billerica, USA), a stand-alone high performance *in vivo/ex vivo* micro-computer tomographe. The lead wire inside the microdialysis membranes is detected with a precision of 9 μm. The Bruker DataViewer™ by Bruker Corporation (Billerica, USA) was used to perform the analysis of the depth measurement.

Figure 12: μ-CT image of a microdialysis experiment with an application site of 1 cm². The membrane depth is shown in inverse color, with the membrane (M) in purple.

5.3.3.2 Measurement via ultrasound

The depth of the linear probe was determined with the Vevo 3100 from FujiFilm Visualsonics (Toronto, Canada). The skin was covered with Aquasonic 100 gel from Parker Laboratories (Fairfield, USA) and the lead wire was detected with the MX700 ultrasound transducer from FujiFilm Visualsonics (Toronto, Canada) inside the skin with a precision of 13 µm. The Vevo LAB3.1.1 software from FujiFilm Visualsonics (Toronto, Canada) was used to perform the analysis of the depth measurement.

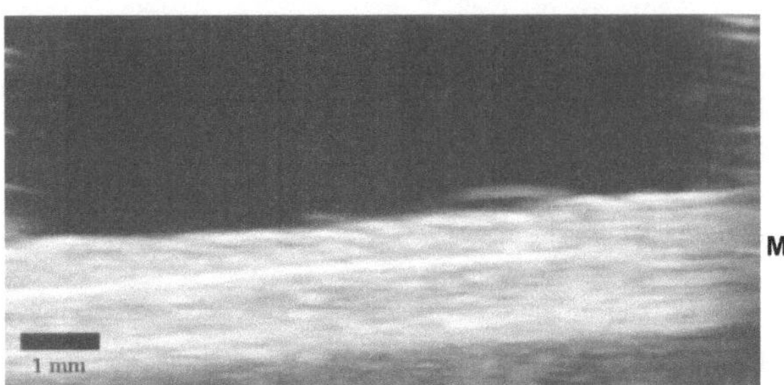

Figure 13: Ultrasound image of a microdialysis experiment with an application site of 1 cm². The membrane depth is shown in black white image, with the membrane (M) in white.

5.3.3.3 Measurement via microscope

For the depth determination images of the MD probes the porcine skin tissue was separated from the cartilage and the application area was cut into 0.5 x 1 cm stripes. The tissue samples were embedded in paraffin and the tissue sectioning and staining protocol was used as described in 5.2.8.

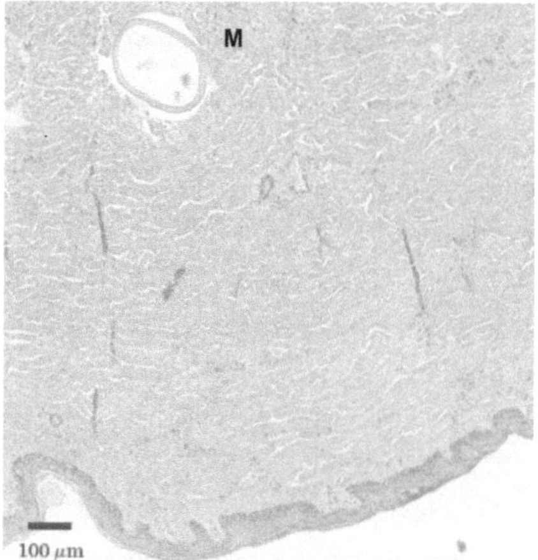

Figure 14: HE image of a microdialysis experiment with an application site of 1 cm². The membrane depth is shown as a HE image, with the membrane (M) core.

5.3.4 Absolute recovery (AR)

The maximum absorption capacity of the unbound active compound to the microdialysis membranes was determined as *in vitro* absolute recovery. The AR was performed at the experimental flow rates of the microdialysis membrane experiments with 3 µL/min. The experiments were performed by inserting the membrane inside of a universal tube. The membranes were perfused with PBS as perfusion fluid. Samples were collected every 30 min for 2 h and analyzed by HPLC. For each time point, the AR was defined as ratio of the concentration of active in the dialysate and the concentration of active in the surrounding medium.

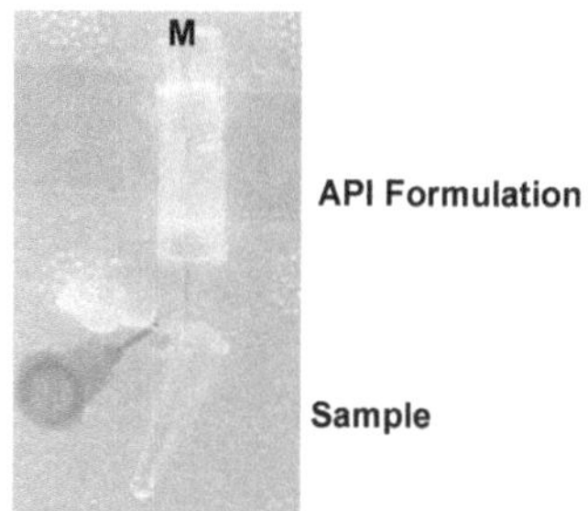

Figure 15: Experimental setup to determine the absoluté recovery. The membrane (M) is placed inside the universal tube containing the API formulation which and the API is collected in the sampling tube (Sample).

$$AR_{(in\ vitro)} = \frac{\left(C_{dialysate} - C_{perfusate}\right)}{\left(C_{medium} - C_{perfusate}\right)} * 100$$

as $C_{perfusate} = 0$

$$AR_{(in\ vitro)} = \frac{\left(C_{dialysate}\right)}{\left(C_{medium}\right)} * 100$$

$AR_{in\ vitro}$ represents the absolute recovery *in vitro*; $C_{dialysate}$ represents the active concentration in the dialysate; and C_{medium} represents the active concentration in the surrounding medium.

5.3.5   Skin recovery (SR)

The maximum absorption capacity of the membrane inside the skin was tested with a formulation containing the active injected close to the membrane beneath the skin to float the skin with the formulation. The skin or *ex vivo* recovery was performed at the experimental flow rates of the microdialysis membrane experiments with 3 µL/min. The membranes were perfused with PBS as perfusion fluid. Samples were collected every 30 min for 2 h and analyzed by HPLC. For each time point, the SR was defined as ratio of the concentration of active in the dialysate and the concentration of active in the surrounding medium.

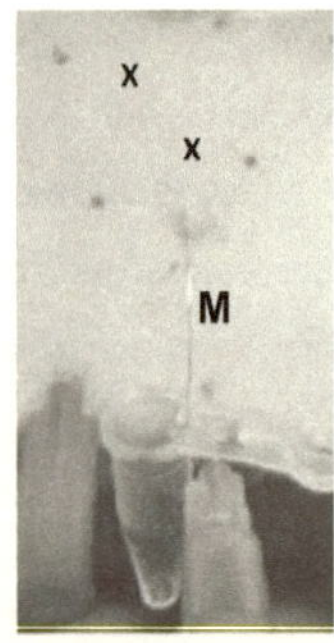

Figure 16: Experimental setup of the SR with the injection sites (X) and the membrane inside the tissue (M), which collects the API inside the sampling tube (Sample).

$$SR_{(ex\ vivo)} = \frac{\left(C_{dialysate} - C_{perfusate}\right)}{\left(C_{medium} - C_{perfusate}\right)} * 100$$

as $C_{perfusate} = 0$

$$SR_{(ex\ vivo)} = \frac{\left(C_{dialysate}\right)}{\left(C_{medium}\right)} * 100$$

$SR_{in\ vitro}$ represented the recovery *ex vivo*; $C_{dialysate}$ represented the active concentration in the dialysate; and C_{medium} represented the active concentration in the surrounding medium.

5.4 Static lateral penetration setup

For the lateral penetration setup, the porcine ear was pinned dermal side down to a solid surface and the hair was cut manually. The application site was defined by a 1 cm² window color fence and 10 µL/cm² of caffeine or LIP1 in a PGOA formulation were topically applied and penetrated the skin for 4 h. The tissue was removed from the cartilage and a 4 cm² with the application site in the center was punched out. From the 4 cm² tissue the 1 cm² application site was punched out. The E and D were separated like described in 5.2.7.

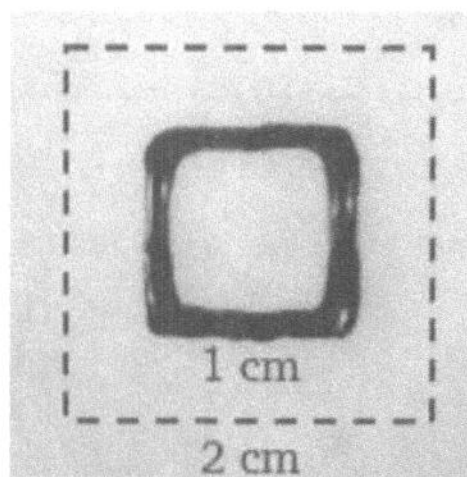

Figure 17: Experimental setup with 1 cm² application site (solid) and the 4 cm² lateral (dotted) penetration site.

5.5 High-Performance Liquid Chromatography (HPLC)

The caffeine and LIP1 concentration were determined using HPLC (VWR-Hitachi ELITE LaChrom system). The quantity of caffeine was detected at a wavelength of 272 nm and for LIP1 at 285 nm with a DAD I-2450 detection unit and a column temperature of 30°C. The analytical determination was performed using a Chromolith ® Performance RP-18e 100-4.6 mm (Merck KGaA, Darmstadt) column as the stationary phase. The results for caffeine were obtained with a flow rate of 2.0 mL/min and an isochratic method with a mobile phase of 90 % water (Milli-Q®) and 10 % acetonitrile (HPLC gradient grade, Merck KGaA, Darmstadt). For LIP1, a flow rate of 2.0 mL/min for the isochratic method with 80 % water and 20 % acetonitrile was used.

Prior to the analysis, the samples were mixed and transferred into auto sampler screw micro-vials (VWR) and analyzed with an injection volume of 10-90 µL of each sample. The chemical stability was confirmed by analyzing the caffeine samples after 7 days at latest. The specificity of the HPLC run was controlled with a blank injection as an internal standard.

5.6 HPLC data analysis

The flux was calculated by the area under the curve from HPLC analysis data with the slope of the calibration curve and with respect to the volume of the receptor chamber and the applied membrane area of each diffusion cell for the FDC setup. For the MD setup, the perfusate and amount (in µg) of active ingredient were detected.

5.7 Calculation

The drug amounts of all samples were calculated as follows. The quantification linearity for the active ingredient was confirmed by measuring a dilution series of an active standard solution of 0.5-250.0 µg/mL. A graphic linear regression conformation was determined by monitoring the regression factor, which was found to be R2 > 0.99 in all cases. Accuracy and precision of the HPLC run was determined within the acceptance criteria of a variation of less than 2 %.

$y = mx + b$ with b = 0

Resulting in $A = mc$

(A = area under the curve [AU], m = slope, c = concentration [µg/mL])

Calculating the concentration with: $c = \frac{A}{m}$ [µg/mL]

And further the total mass of penetrated active: $m = c * V$ [µg]

(V = volume in which the active was solved)

Using the application area of each FDC, the cumulative amount Q of penetrated drug can be calculated:

$Q = \frac{c}{A_{FDC}}$ [µg/cm²]

The drug amount per mass of skin layer, and for the finite experiments the amount of drug as a percentage of the applied amount, were determined.

All calculations were made using the software program Microsoft® Excel® Office 365 and GraphPad Prism 8.03.

5.8 **Statistical analysis**

Statistical analysis was performed using GraphPad Prism 7.03 (GraphPad Software, San Diego, USA). All the data are shown as the mean ± SD. The statistical analyses were performed using t-tests or one-way ANOVA statistics (the Bonferroni method for multiple comparisons was also applied). Probability (p) values less than 0.05 were considered statistically significant.

6 Results

For the penetration abilities of APIs, caffeine and LIP1 were tested for their target-site penetration into skin. These APIs were chosen since they have different physico-chemical characteristics and are tested in an FDC and MD setup in finite and infinite dosage forms using different formulations.

6.1 Validation of the FDC system

In the experimental percutaneous skin penetration setup, the influences of different human skin surrogates, changes to the fluid in the donor chamber, and different receptor chamber conditions are tested. Influences on the penetration of the model active, caffeine, for topical application are shown. The standard penetration flux was determined for an infinite (786 µL/cm²) dosage of 0.7 % caffeine formulation on different skin surrogates. For *ex vivo* human and porcine split-skin penetration studies of 500 µm thick split-skin was used.

6.1.1 Human skin surrogate comparison

To study skin penetration, *ex vivo* human and porcine split-skin was used. For the *in vitro* penetration study, the artificial membrane, Strat-M˚, was used as a surrogate for human skin.

6.1.1.1 Validation of the skin surrogate and determine the effect of the operator

Skin thickness on penetration testing in the FDC

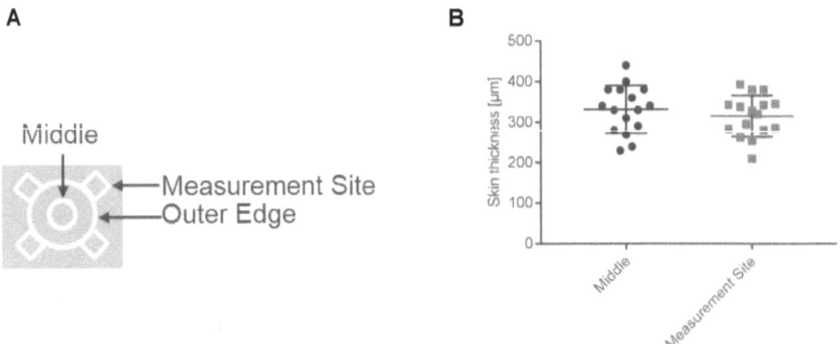

Figure 18: Schematic illustration of the skin disc thickness measurement (A). Determination of the thickness of dermatomed skin discs (B). Comparison of micrometer measurements of skin determined four peripheral measurement sites (blue) vs a central measurement site (black). Values represent mean (n = 16) ± SD, compared using t-Test statistics.

Figure 18 shows the comparability of the thickness of dermatomized skin by the measurement of four sites of the skin disc and the middle of the disc. The mean measurements of the middle with 330 ± 60 µm had no statistically significant differences than 315 ± 55 µm for the measurement sites.

Table 4: Defined dermatomized thickness of skin disc and mean thickness of each skin layer.

Thickness setting on dermatome [µm]	Skin layer	Measured thickness of skin layer [µm]
300	SC	12.05 ± 2.70
	E	101.6 ± 34.2
	D	180.4 ± 86.4
500	SC	12.3 ± 3.2
	E	105.3 ± 27.5
	D	488.3 ± 110.9
700	SC	9.9 ± 0.8
	E	111.8 ± 24.6
	D	702.5 ± 99.20
1000	SC	14.6 ± 3.9
	E	124.3 ± 46.7
	D	970.8 ± 203.9

The defined dermatome thickness setting of 300, 500, 700 and 1000 µm was tested and the mean thickness of each skin layer was determined. For each defined thickness the mean SC was between 9.1 µm and 18.5 µm and the E between 77.6 µm and 171 µm thick. The D layer was progressively thicker from the thinnest dermatomed skin disc to the thickest, ranging from 94 to 1174.7 µm. Split-skin comprises always the SC and E plus an increasing thickness of the D depending of the dermatomed split-skin thickness.

To validate the integrity of the different skin layers the TEWL was measured.

TEWL integrity determination

To determine the integrity of the skin surrogates the TEWL was measured for the Stat-M membrane, the porcine and human split-skin.

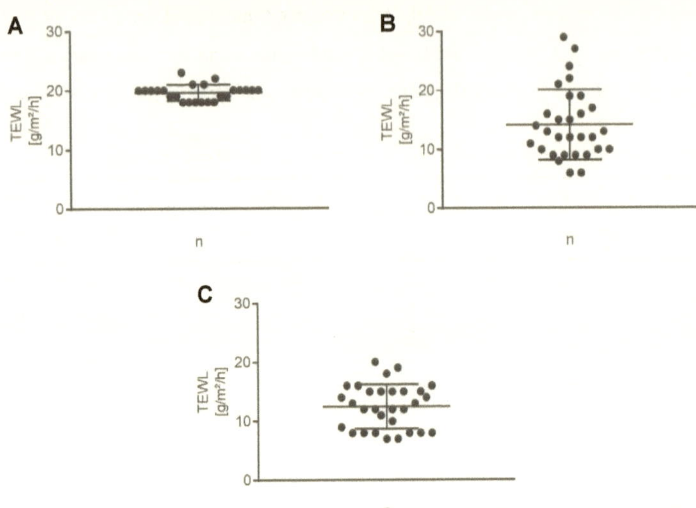

Figure 19: TEWL of different human skin surrogates. The TEWL of the Strat-M® membrane (A) (n = 24), 500 µm porcine split-skin (B) (n = 30) and 500 µm human split-skin (C) (n = 30) mounted on a FDC was determined with a TEWL device. Values represent mean ± SD.

In Figure 19 the mean TEWL of the Strat-M® membrane, porcine, and human split-skin were determined to be 20 ± 1, 14 ± 6 and 12.5 ± 3.5 g/m²/h, respectively. Therefore, the SD for the TEWL measurements was Strat-M®< porcine skin< human skin.

The influence of the thickness on the TEWL of the skin was measured for 300 to 1000 µm thick porcine split-skin.

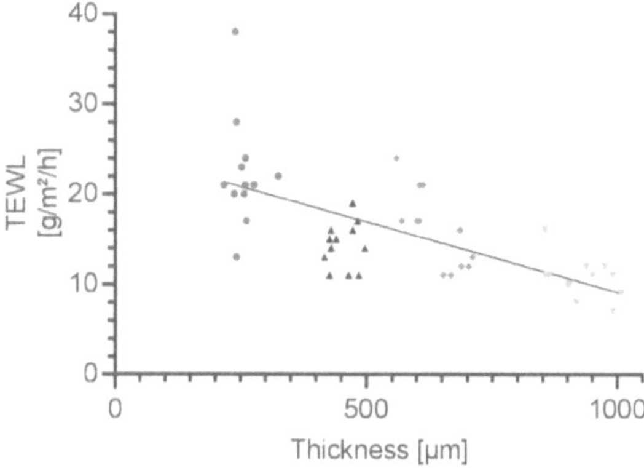

Figure 20: Correlation of TEWL vs skin thickness. Measured TEWL values are plotted against the measured thickness of porcine split-skin obtained using the dermatome setting 300 µm (blue), 500 µm (black), 700 µm (grey) and 1000 µm (light grey). m = -0.0156, R^2 = 0.4589 (n = 12).

Figure 20 shows the TEWL values from dermatomed porcine split-skin obtained at settings of 300, 500, 700 and 1000 µm which was measured for its thickness. With an increase in skin thickness, lower TEWL values and a downward trajectory of the linear regression is observed. Similarly, thicker sections appear to show a lower SD than thinner sections.

Due to biological variations from donor to donor and experiment to experiment the variations range was determined.

6.1.1.2 Experimental variation of the *in vitro* and *ex vivo* penetration testing of caffeine

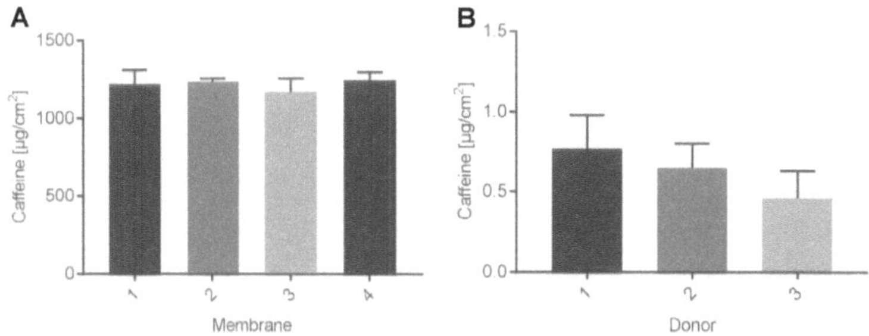

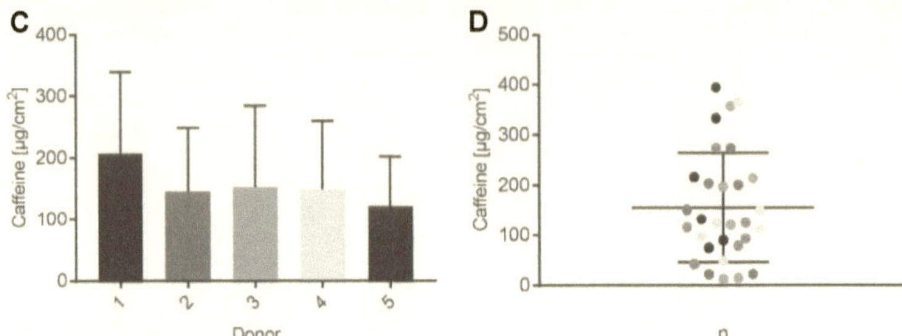

Figure 21: Experimental variation of different human skin surrogates. The experimental variation and the caffeine flux through the Strat-M® membrane (A), 500 µm human split-skin (B) and porcine split-skin (B) mounted on the FDC was determined. Each bar represents the caffeine penetration flux for six replicates (n = 6) from a single donor for 4 h at 32°C and a topical application of 786 µL 0.7 % caffeine PGOA formulation. The caffeine flux through the porcine split-skin shows the inter (C) and intra (D) biological variance. Values represent mean ± SD, compared using one-way ANOVA statistics.

Figure 21 illustrates the biological variation of different skin surrogates, where 1172 to 1247 µg/cm² caffeine penetrated through the Strat-M® membrane. 121 to 207 µg/cm² caffeine penetrated through the porcine skin and 0.35 to 0.77 µg/cm² through the human skin. All similar surrogates show no statistically significant difference for the experiments. The SD regarding the measured penetration experiments was lowest for Strat-M®, intermediate for human skin and highest for porcine skin.

6.1.1.3 Impact of kinetic sampling on absolute caffeine concentration in the FDC

The impact of different sampling techniques, of individual experiments per time point and cumulative sampling for 1 h and 2 h was determined for kinetic penetration experiments on different human surrogates.

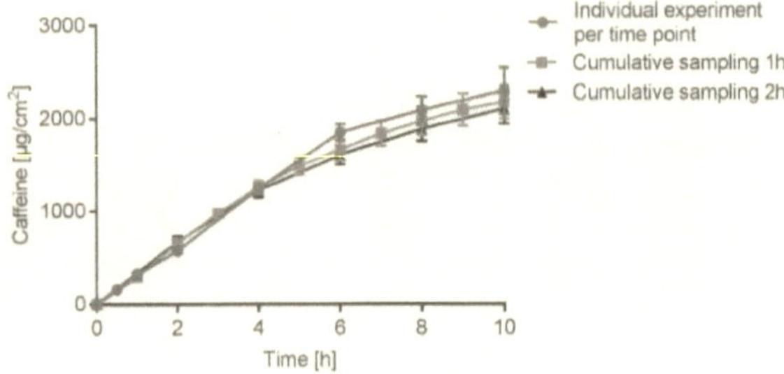

Figure 22: Kinetic caffeine flux through the Strat-M® membrane. The caffeine flux through the Strat-M® membrane for 10 h at 32°C for a topical application of 786 µL 0.7 % caffeine PGOA formulation was determined. Within the blue line every time point represents the caffeine penetration flux for six replicates (n = 6) of individual experiments. The grey and black line represents the caffeine penetration flux for six replicates (n = 6) for one experiment with 1 h and 2 h cumulative sampling, respectively. Values represent mean ± SD, compared using one-way ANOVA statistics.

Figure 22 shows the kinetic caffeine flux through the Strat-M® membrane. The experiments show a similar kinetic with no statistically significant differences regardless if single experiments or cumulative sampling for 1 h or 2 h were performed. Over 10 h a hyperbolic curve with a caffeine maximum of 2105 to 2296 µg/cm² with a range difference of 191 µg/cm², which is 12 % of the maximum, was observed. After 4 h the caffeine flux of 1230 to 1261 µg/cm² with a range difference of 31 µg/cm², which is 2 %, was observed. The maximum amount of caffeine that penetrated the Strat-M® membrane after 4 h is 55 % of the maximum after 10 h.

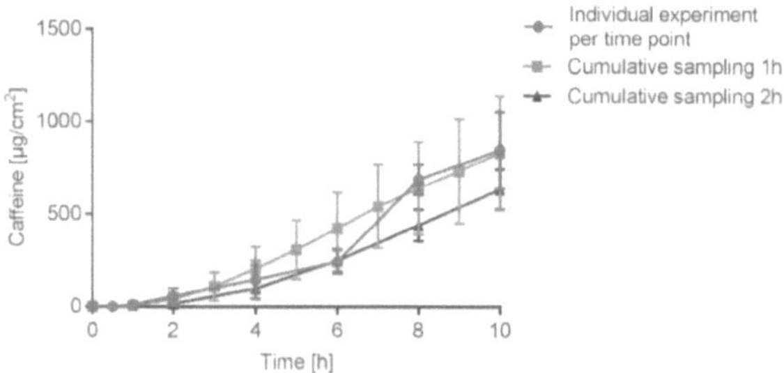

Figure 23: Kinetic caffeine flux through porcine skin. The caffeine flux through 500 µm porcine split-skin for 10 h at 32°C and a topical application of 786 µL 0.7 % caffeine PGOA formulation was determined. Within the blue line every time point represents the caffeine penetration flux for six replicates (n = 6) of individual experiments. The grey and black line represents the caffeine penetration flux for six replicates (n = 6) for one experiment with 1 h and 2 h cumulative sampling, respectively. Values represent mean ± SD, compared using one-way ANOVA statistics.

Figure 23 shows the kinetic caffeine flux through 500 µm porcine split-skin. The experiments show a similar kinetics and no statistically significant differences were found between independent single sampling or cumulative sampling for 1 h or 2 h. Over 10 h a hyperbolic curve with a caffeine maximum of 636 to 830 µg/cm² with a range difference of 194 µg/cm², which is 24 % of the maximum, was observed. After 4 h the caffeine flux of 98 to 204 µg/cm² with a range difference of 106 µg/cm², which is 52 %, was observed. The maximum amount of caffeine that penetrated the porcine skin after 4 h is 25 % of the maximum after 10 h.

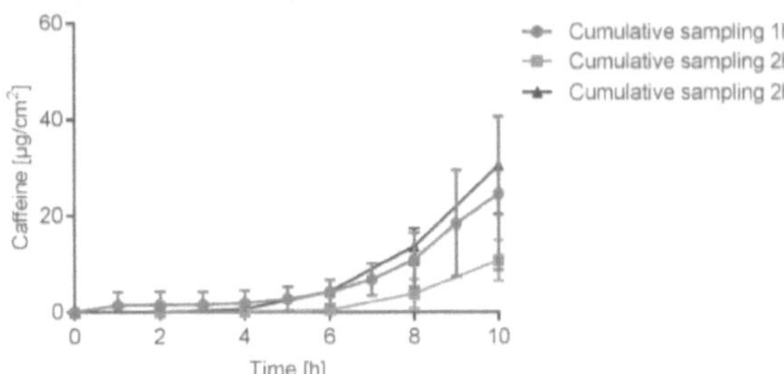

Figure 24: Kinetic caffeine flux through human skin. The caffeine flux through 500 µm human split-skin for 10 h at 32°C and a topical application of 786 µL 0.7 % caffeine PGOA formulation was determined. The blue line represents the caffeine penetration flux for six replicates (n = 6) for one experiment with 1 h cumulative sampling. The grey and black line represent the caffeine penetration flux for two donors with six replicates (n = 6) for one experiment with 2 h cumulative sampling. Values represent mean ± SD, compared using one-way ANOVA statistics.

Figure 24 shows the kinetic caffeine flux through 500 µm human split-skin. The experiments show a similar kinetic trend with no statistically significant differences for independent cumulative sampling or cumulative sampling every 1 h or 2 h. Over 10 h a beginning exponential curve with a caffeine maximum of 11 to 31 µg/cm² with a range difference of 20 µg/cm², which is 65 % of the maximum, was observed. After 4 h the caffeine flux of 0.08 to 1.9 µg/cm² with a range difference of 1.82 µg/cm², which is 96 %, was observed. The maximum amount of caffeine that penetrated the human skin after 4 h is 6 % of the maximum after 10 h.

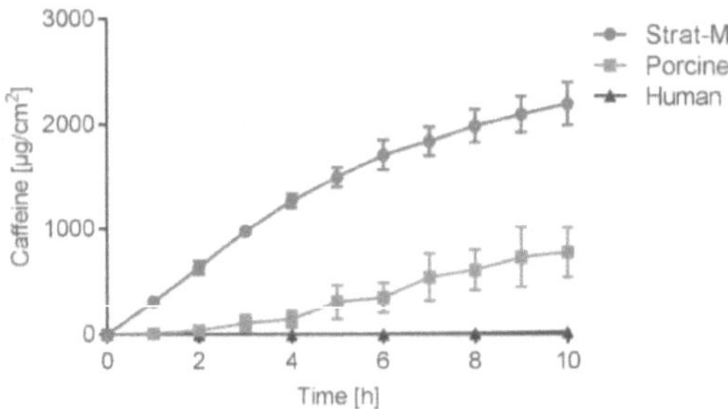

Figure 25: Kinetic caffeine flux through different human skin surrogates. The caffeine flux through Strat-M®, porcine and human split-skin for 10 h at 32°C and a topical application of 786 µL 0.7 % caffeine PGOA formulation was determined. The blue line represents the mean caffeine penetration flux through the Strat-M® membrane (n = 18). The grey line represents the mean caffeine penetration flux for three donors with six replicates (n = 18) through 500 µm porcine split-skin and black line represent the mean caffeine penetration flux for three donors with six replicates (n = 18) for through 500 µm human split-skin. Values represent mean ± SD.

Figure 25 shows caffeine flux over 10 h through the Strat-M® membrane, 500 µm porcine split-skin, and 500 µm human split-skin. The penetration kinetics through the Strat-M® membrane and porcine skin show a hyperbolic curve with a maximum of 2192 µg/cm² and 779 µg/cm² after 10 h. Strat-M® has a 2.8-fold penetration compared to porcine skin. Human skin shows after 10 h a flux of 22 µg/cm² which is 100-fold less than Strat-M® and 35-fold less than porcine skin. For the membrane the half maximum caffeine flux was reached after 3.5 h, for porcine skin after 6.3 h and for human skin after 8.3 h.

6.1.2 Influences of donor chamber conditions

For percutaneous penetration studies, the API vehicle, the API concentration and the amount of applied formulation are essential. The influences of the donor chamber conditions with a change in formulation, volume and concentration were tested for a topical application of caffeine at 32°C for 4h.

6.1.2.1 Effect of the formulation on the penetration

The influences of a 0.7 % caffeine PGOA, PG and H_2O formulation on the penetration through the Strat-M® membrane, porcine and human split-skin were determined.

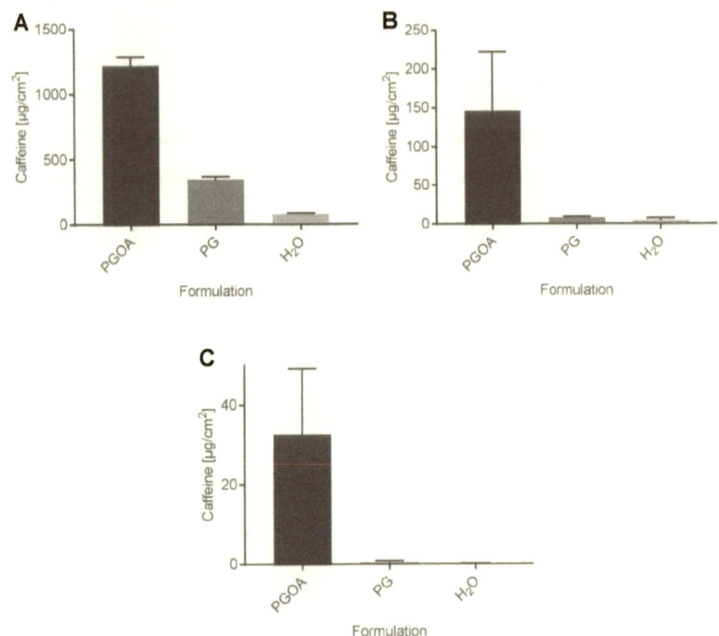

Figure 26: Influences of the formulation on the caffeine flux through different human skin surrogates. The PGOA, PG and H_2O caffeine formulations and the caffeine flux through the Strat-M® membrane (A), 500 µm porcine split-skin (B) and 500 µm human split-skin (C) mounted on the FDC were determined. Each bar represents the caffeine penetration flux for six replicates (n = 6) from a single donor for 4 h at 32°C and a topical application of 786 µL and a 0.7 % caffeine formulation. Values represent mean ± SD.

Figure 26 shows the influences on the penetration of different caffeine formulations on Strat-M® membrane, porcine split-skin, and human split-skin. The penetration and flux of caffeine was for Strat-M® membrane > porcine skin > human skin. In a PGOA formulation caffeine penetrated the Strat-M® membrane 3.6-fold more the than the PG formulation and 15.6-fold more than the water formulation. The concentration of penetrated caffeine with the PGOA resulted in 19.2-fold higher flux than the PG formulation and 33-fold higher compared to the water formulation through porcine skin. For human skin, caffeine penetrated using a PGOA formulation 77.6-fold higher than for the PG formulation and 500-fold compared to the water formulation. The SD of the penetration experiments was for Strat-M® membrane < human skin < porcine skin.

6.1.2.2　Effect of the donor volume on penetration

The influences of a 0.7 % caffeine PGOA formulation topically applied in a finite (<100 µL/cm²) and infinite (>100 µL/cm²) amount on the penetration through the Strat-M® membrane and porcine split-skin was determined.

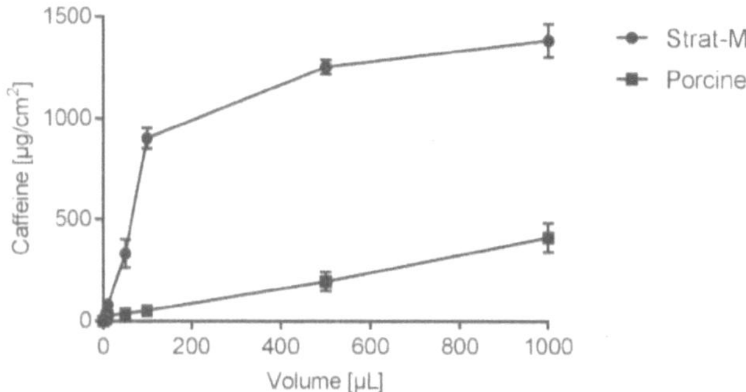

Figure 27: Influences of the volume amount on the caffeine flux through different human skin surrogates. The influences of different volumes and the caffeine flux through the Strat-M® membrane and 500 µm porcine split-skin mounted on the FDC were determined. Each volume represents the caffeine penetration flux for six replicates (n = 6) from a single donor for 4 h at 32 and a topical application of a 0.7 % caffeine PGOA formulation. Values represent mean ± SD.

Figure 27 shows the influences on the caffeine penetration of different formulation volumes on the Strat-M® membrane and porcine split-skin. Caffeine penetrated with a linear flux through the Strat-M® membrane up to a volume of 100 µL and penetrated with a volume > 100 µL in hyperbolic flux until a plateau was reached. For the porcine skin caffeine penetrated with a linear correlation of volume and flux up to 1000 µL. With a topical application of 1000 µL, 1380 µg/cm² caffeine penetrated the Strat-M® membrane and 411 µg/cm² the porcine skin. The SD of the penetration experiments was less for Strat-M® than for porcine skin.

6.1.2.3   Effect of caffeine saturation on penetration

The influences of the saturation of caffeine in a PGOA formulation topically applied on the penetration through the Strat-M® membrane and porcine split-skin was determined.

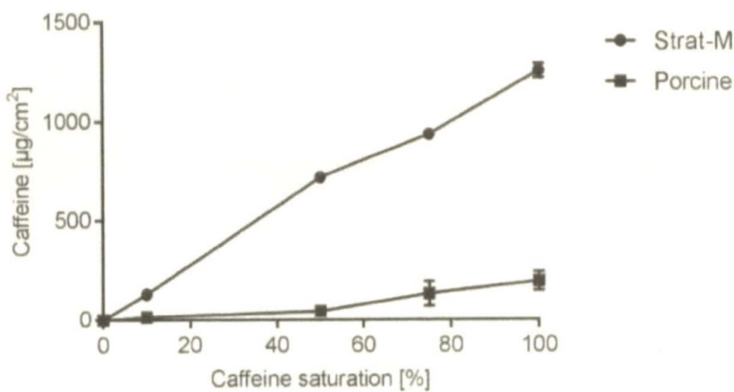

Figure 28: Influences of the caffeine saturation on the caffeine flux through different human skin surrogates. The influences of caffeine saturation and the caffeine flux through the Strat-M® membrane and 500 µm porcine split-skin mounted on the FDC were determined. Each saturation concentration represents the caffeine penetration flux for six replicates (n = 6) from a single donor for 4 h at 32°C and a topical application of 786 µL/cm² of a caffeine PGOA formulation. Values represent mean ± SD.

Figure 28 shows the influences on the caffeine penetration of different caffeine concentrations on the Strat-M® membrane and porcine split-skin. Caffeine penetrated with a linear correlation to the concentration, with 100% of saturation 6.4-fold through the Strat-M® membrane compared to the porcine split-skin. The SD of the penetration experiments was less for Strat-M® compared to porcine skin.

6.1.3 Influences of receptor chamber conditions

To study the percutaneous skin penetration *ex vivo* porcine skin was used. For the *in vitro* penetration study, the artificial Strat-M® membrane was used as a surrogate for human skin. The influences of the receptor chamber with a change in temperature and receptor fluid were tested for a topical application of a saturated caffeine PGOA formulation for 4 h at 32°C.

6.1.3.1 Effect of the temperature on the penetration

The influences of the receptor fluid temperature on the caffeine flux in a PGOA formulation topically applied on the penetration through the Strat-M® membrane and porcine split-skin are determined.

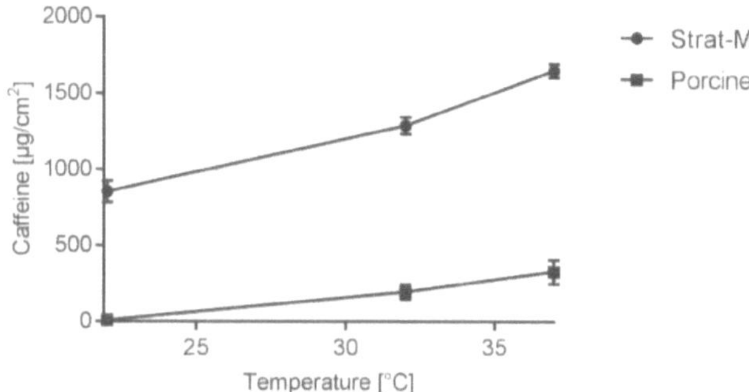

Figure 29: Influences of the receptor fluid temperature on the caffeine flux through different human skin surrogates. The influence of the receptor chamber temperature on the caffeine flux through the Strat-M® membrane and 500 µm porcine split-skin mounted on the FDC were determined. Each temperature represents the caffeine penetration flux for six replicates (n = 6) from a single donor for 4 h at 32°C and a topical application of 786 µL/cm² of a 0.7 % caffeine PGOA formulation. Values represent mean ± SD.

Figure 29 shows the influences on the caffeine penetration of different receptor chamber temperatures on the Strat-M® membrane and porcine split-skin. Caffeine penetrated with a linear flux 5-fold more through the Strat-M® membrane than through porcine skin. At 37°C 1650 µg/cm² caffeine penetrated through the Strat-M® membrane and 330 µg/cm² through porcine skin. The SD of the penetration experiments was less for Strat-M® than for porcine skin.

6.1.3.2 Effect of the receptor fluid on the penetration

The influences of multiple receptor fluids which are used for lipophilic APIs were tested on their penetration influences for the caffeine flux in a PGOA formulation topically applied through the Strat-M® membrane and porcine split-skin.

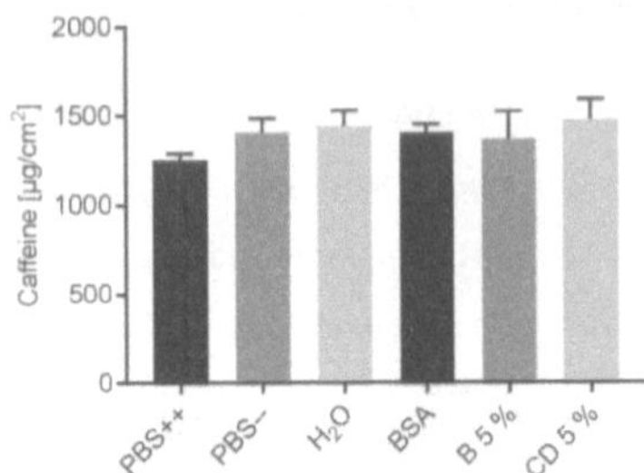

Figure 30: Influences of the receptor fluid on the caffeine flux through the Strat-M® membrane. The influence of the receptor fluid on the caffeine flux through the Strat-M® membrane mounted on the FDC were determined. Each bar represents the caffeine penetration flux for six replicates (n = 6) for 4 h at 32°C and a topical application of 786 µL/cm² of a 0.7 % caffeine PGOA formulation. With PBS++ (with Mg^{2+} and Ca^{2+}), PBS-- (without Mg^{2+} and Ca^{2+}), H_2O (Milli-Q), BSA (1 % BSA and 0.9 % NaCl in H_2O), B 5 % (Brij 5 % in H_2O) and CD 5 % (Cyclodextrin 5 % in PBS++) in the receptor chamber. Values represent mean ± SD, compared using one-way ANOVA statistics.

Figure 30 shows the influence on the caffeine penetration through the Strat-M® membrane for different receptor fluids. Between 1253 µg/cm² and 1477 µg/cm² caffeine penetrated through the membrane, therefore PBS++ showed the lowest and 5 % Cyclodextrin in PBS++ the highest penetration flux as a receptor fluid. The receptor fluids PBS--, H_2O, BSA, B 5 % and CD 5 % show no statistically significant differences.

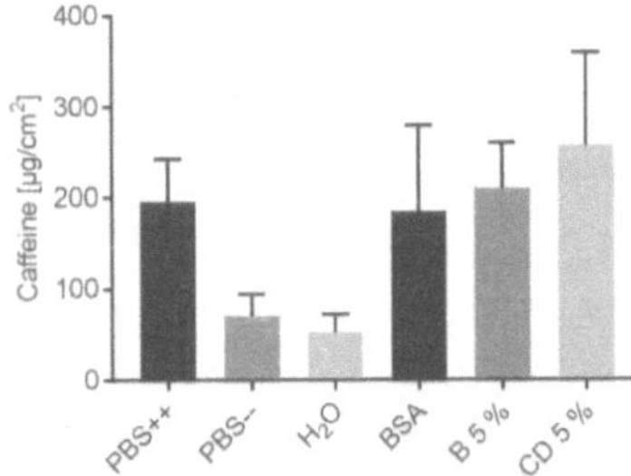

Figure 31: Influences of the receptor fluid on the caffeine flux through the 500 µm porcine split-skin. The influence of the receptor fluid on the caffeine flux through 500 µm porcine split-skin mounted on the FDC were determined. Each bar represents the caffeine penetration flux for six replicates (n = 6) from a single donor for 4 h at 32°C and a topical application of 786 µL/cm² of a sat. caffeine PGOA formulation. With PBS++ (with Mg^{2+} and Ca^{2+}), PBS-- (without Mg^{2+} and Ca^{2+}), H_2O (Milli-Q), BSA (1 % BSA and 0.9 % NaCl in H_2O), B 5 % (Brij 5 % in H_2O) and CD 5 % (Cyclodextrin 5 % in PBS++) in the receptor chamber. Values represent mean ± SD, compared using one-way ANOVA statistics.

Figure 31 shows the influence on the caffeine penetration through porcine skin for different receptor fluids. PBS-- and H_2O show with 52 µg/cm² and 71 µg/cm² the lowest and therefore the baseline amount of penetrated caffeine 5 % CD in PBS shows with 257 µg/cm² the highest penetration flux. The receptor fluids PBS++, BSA, B 5 % and CD 5 % show no statistically significant differences.

 Influences of the API lipophilicity on penetration

For the experimental skin penetration setup and an understanding of the influences of molecular weight and lipophilicity *ex vivo* porcine skin was used. Caffeine as a hydrophilic active and the lipophilic active LIP1 have the identical molecular weight and are tested for their penetration ability in different skin layers. The influences on the penetration for topical application of a finite (10 µL/cm²) and infinite (786 µL/cm²) amount with 0.7 % active in a PG and PG with 5 % OA formulation were tested.

6.2.1 Skin penetration using the FDC setup

For the *ex vivo* percutaneous penetration studies 500 µm porcine split-skin was mounted on the FDC filled with PBS++ at 32°C. Therefore, the penetration features into the SC, viable epidermis (E) and dermis (D) of caffeine and LIP1 for a PG and PGOA formulation were tested as well as their ability of lateral penetration.

6.2.1.1 Effect of the skin thickness on the penetration

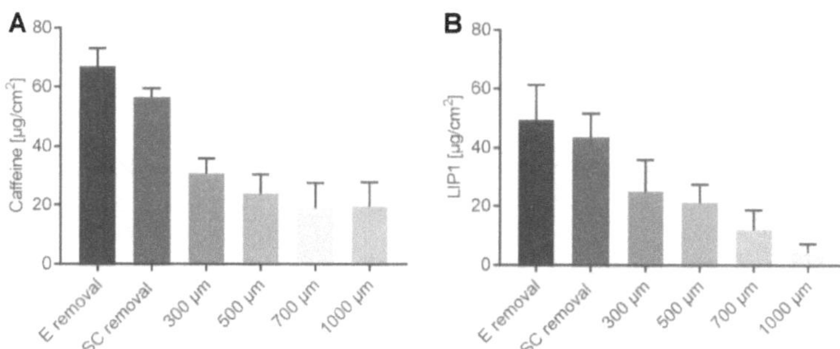

Figure 32: Influences of the different skin layers on the caffeine flux through porcine split-skin. The influences of the E and SC removal of 500 µm split-skin and different split-skin thicknesses on the caffeine (A) and LIP1 (B) flux through porcine split-skin mounted on the FDC were determined. Each bar represents the caffeine or LIP1 penetration flux for six replicates (n = 6) from a single donor for 4 h at 32°C and a topical application of 10 µL/cm² of a caffeine or LIP1 PGOA formulation. Values represent mean ± SD, compared using t-Test statistics.

Figure 32 shows the influences on the caffeine and LIP1 penetration for different split-skin thicknesses. Additionally, the removal of the E and SC on 500 µm thick porcine split-skin was tested for the penetration ability of caffeine. Caffeine penetrated the skin with the lowest flux of 19 µg/cm² up to 67 µg/cm². Depending on the D thickness, between 19 to 31 µg/cm² caffeine penetrated the skin with a low gradient. The removal of the E in contrast to the SC removal increased the flux 1.18-fold. Due to the SC removal 1.84-fold more caffeine penetrated through the skin which is an increase of 84 % on the penetration. LIP1 penetrated the skin with the lowest flux of 4 µg/cm² up to 49 µg/cm². Depending on the D thickness, between 4 to 25 µg/cm² LIP1 penetrated the skin with a high gradient. The removal of the E in contrast to the SC removal increased the flux 1.12-fold. Due to the SC removal 1.74-fold more LIP1 penetrated through the skin which is an increase of 74 % on the penetration. The removal of the lipophilic SC had a statistically significant influence on the penetration on the hydrophilic caffeine but not for lipophilic LIP1. The influences of the different skin layers on the penetration of caffeine and LIP1 are shown in Figure 32 but the understanding of the total amount of API in the different layers and amount per skin layer weight is missing.

6.2.1.2   Skin layer dependent penetration

For the experimental percutaneous skin penetration setup, the influences of different skin layers (SC, E and D) on the lipophilicity of the API were tested. The influences of the formulation (PG/PGOA), application volume (10/786 µL/cm²) and kinetics are shown. For the *ex vivo* penetration study 500 µm porcine split-skin was used and experiments with a total recovery ≥ 70 % are shown in percentage of total active determined. Therefore, the amount of the active found in the different layers is defined as lateral (Lateral), SC, E, and D penetration as well as active which did not penetrate (SW) and active which penetrated the whole skin inside the receptor fluid (RF).

6.2.1.3   Penetration kinetics for a finite applied volume of different formulations

The influences of the lipophilicity of caffeine and LIP1 for finite application of PG and PGOA formulation were tested.

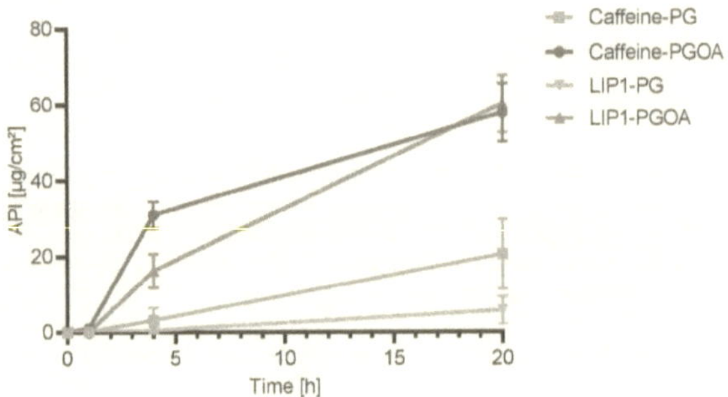

Figure 33: Influences of the API and the formulation on the penetration. The kinetic penetration of caffeine (blue) and LIP1 (red) through 500 µm porcine split-skin for 1 h, 4 h and 20 h at 32°C was determined. Every time point represents the penetration flux for six replicates (n = 6) from a single donor for a topical application of 10 µL/cm² of 0.7 % active in a PG (lighter) or PGOA (darker) formulation. Values represent mean ± SD.

Figure 33 shows the caffeine and LIP1 penetration flux for a finite topical application of a PG and PGOA formulation over time through porcine skin. The penetration of the active into the skin was increased using PG with OA. Caffeine penetrated the skin more efficiently compared to LIP1 for the PG and PGOA formulation, except for the PGOA formulation after 20 h. After 4 h caffeine penetrated within the PGOA formulation the skin with 31 µg/cm² 1.9-fold and after 20 h 0.96-fold of LIP1. For the PG formulation, caffeine penetrated the skin after 20 h 2.8-fold more and LIP1 10.5-fold more.

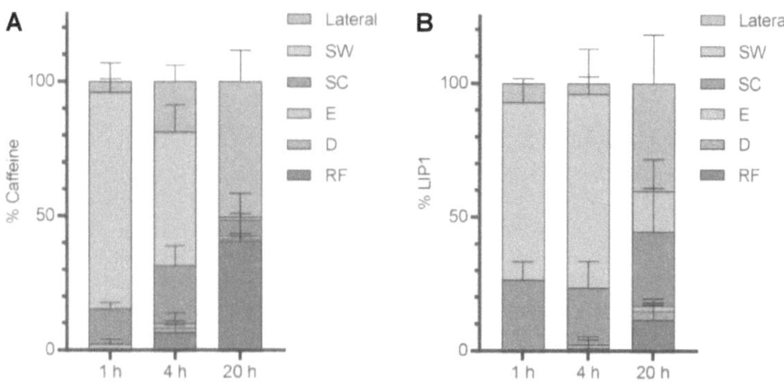

Figure 34: Influences of the API in a PG formulation on the penetration. The mean distribution of caffeine (A) and LIP1 (B) in the SC, E and D of 500 µm porcine split-skin after 1 h, 4 h and 20 h at 32°C was determined. A finite dose of 10 µL/cm² of 0.7 % active solved in a PG formulation was topically applied. Each bar shows the percentage of determined active in each layer (Lateral, SW, SC, E, D and RF) and represents the penetration for six replicates (n = 6) from a single donor. Values represent mean ± SD and a total recovery of the API > 70 % of the application amount.

Figure 34 shows the penetration of caffeine and LIP1 for a finite topical application of a PG formulation. The figure shows an increase of caffeine and LIP1 in the RF and a decrease in the SW over time with a higher amount and faster penetration for caffeine. After 4 h and 20 h, 7 % and 40 % of determined caffeine and after 20 h 13 % of the determined LIP1 penetrated into the RF. Caffeine penetrates the skin in deeper skin layers (E and D) faster and with a higher amount than LIP1. The lateral penetration increases over time for both actives and especially caffeine shows with 50 % of the determined active a higher lateral penetration after 20 h. For LIP1 the SC is the main reservoir of the different skin layers and for every time point the highest amount of determined active inside the skin, after 1 h 27 %, after 4 h 23 % and after 20 h 26 %.

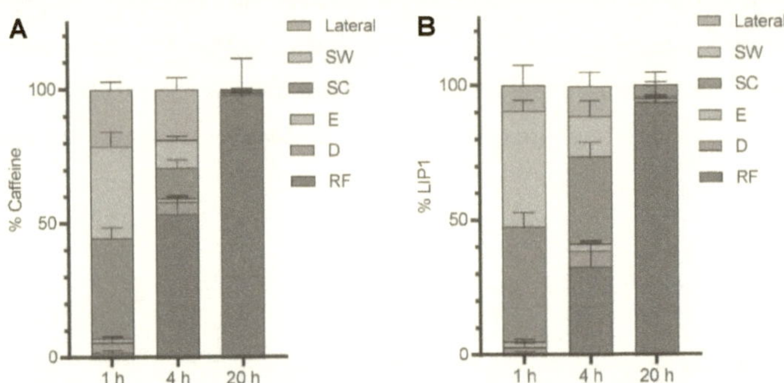

Figure 35: Influences of the API in a PGOA formulation on the penetration. The mean distribution of caffeine (A) and LIP1 (B) within the SC, E and D of 500 µm porcine split-skin after 1 h, 4 h and 20 h at 32°C was determined. A finite dose of 10 µL/cm² of 0.7 % active solved in a PGOA formulation was topically applied. Each bar shows the percentage of determined active in each layer (Lateral, SW, SC, E, D and RF) and represents the penetration for six replicates (n = 6) from a single donor. Values represent mean ± SD and a total recovery of the API > 70 % of the application amount.

Figure 35 shows the penetration of caffeine and LIP1 for a finite topical application of a PGOA formulation. The figure shows an increase of caffeine and LIP1 in the RF and a decrease in the SW over time with a higher amount and faster penetration for caffeine. After 20 h 98 % of determined caffeine and 93 % of determined LIP1 was located in the RF. Caffeine penetrates the skin in deeper skin layers (E and D) within 1 h with a higher amount than LIP1. The total determined API which penetrated lateral increases over time for both actives but remains constant with ~20 % of the determined caffeine and with 10-12 % of the determined LIP1 for 1 h and 4 h experiments. For LIP1 the SC is the main reservoir of the different skin layers and for every time point the highest amount of determined active inside the skin, after 1 h 36 %, after 4 h 35 % and after 20 h 5 %.

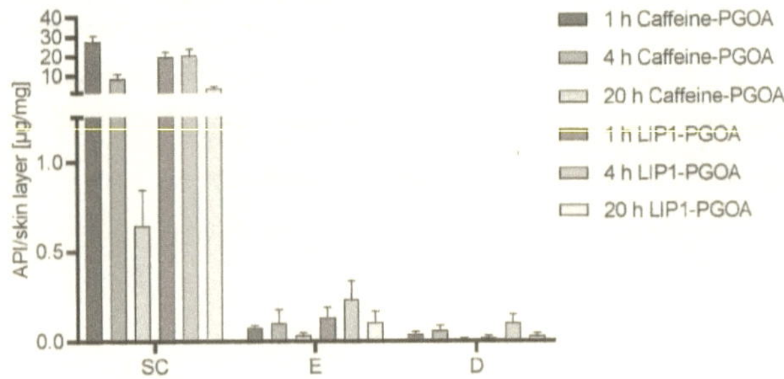

Figure 36: The mean amount of caffeine (blue) and LIP1 (red) per skin layer within the SC, E and D of 500 µm porcine split-skin after 1 h, 4 h and 20 h at 32°C was determined. A finite dose of 10 µL/cm² of 0.7 % active solved in a PGOA formulation was topically applied. Each bar shows the amount of mass per skin layer (SC, E and D) and represents the penetration for six replicates (n = 6) from a single donor. Values represent mean ± SD and a total recovery of the API > 70 % of the application amount.

Figure 36 shows the mean amount of caffeine and LIP1 which penetrated in the different skin layers for a finite topical application of a PGOA formulation for a kinetic of 1 h, 4 h and 20 h. The SC, E and D layers show for each time point the similar increase and decrease trend and an equilibrium distribution for each layer of the total determined active amount. Except for the amount of caffeine and LIP1 in the SC after 1 h, there is a higher amount than for other time points and skin layer correlations. The amount of caffeine and LIP1 per skin layer is for the SC< E< D for every time point.

6.2.1.4 Penetration differences for biological variation for a finite formulation volume

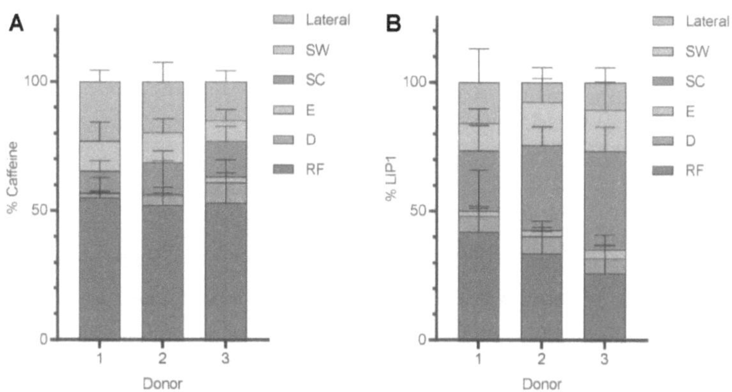

Figure 37: The mean distribution of caffeine (A) and LIP1 (B) within the SC, E and D of 500 µm porcine split-skin for three different donors was determined. A finite dose of 10 µL/cm² of 0.7 % active solved in a PGOA formulation was topically applied for 4h at 32°C. Each bar shows the percentage of determined active in each layer (Lateral, SW, SC, E, D and RF) and represents the penetration for six replicates (n = 6) from a single donor. Values represent mean ± SD and a total recovery of the API > 70 % of the application amount.

Figure 37 shows the mean distribution of caffeine and LIP1 within the different layers of skin for a finite topical application of a PGOA formulation for three donors. For all three donors the distribution of caffeine inside the RF is with > 50 % higher than for LIP1 with 38-42 %. The skin layers show a higher distribution of caffeine inside the SC> D> E, this is similar for LIP1 except that the SC shows with 23-39 % of determined LIP1 a higher amount of active. LIP1 shows for each donor 2-3 % of active inside the E and 6-8 % inside the D layer. The receptor chamber shows for the LIP1 penetration for donor one with 42 % the highest amount of active for donor two and three the SC exhibits more active than the 26-34 % of RF. The lateral penetration has with 17-23 % a higher distribution for the penetration of caffeine than with 7-15 % for LIP1 for all three donors.

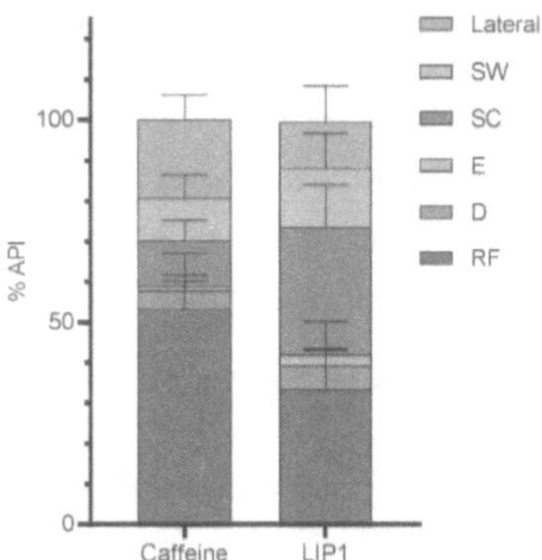

Figure 38: The mean of the distribution of caffeine and LIP1 within the SC, E and D of 500 µm porcine split-skin porcine for three different donors was determined. A finite dose of 10 µL/cm² of 0.7 % active solved in a PGOA formulation was topically applied for 4 h at 32°C. Each bar shows the percentage of determined active in each layer (Lateral, SW, SC, E, D and RF) and represents the penetration for six replicates (n = 6) from a single donor. Values represent mean ± SD and a total recovery of the API > 70 % of the application amount.

Figure 38 shows the mean of the distribution of caffeine and LIP1 within the different layers of skin for a finite topical application of a PGOA formulation for three donors. After a penetration for 4 h caffeine was prominently determined inside the RF> Lateral> SC> SW> E> D. LIP1 follows the distribution of SC> RF> SW> Lateral> D> E. Inside the RF the caffeine amount is 1.6-fold higher than LIP1 and inside the SC LIP1 is 2.8-fold higher than caffeine.

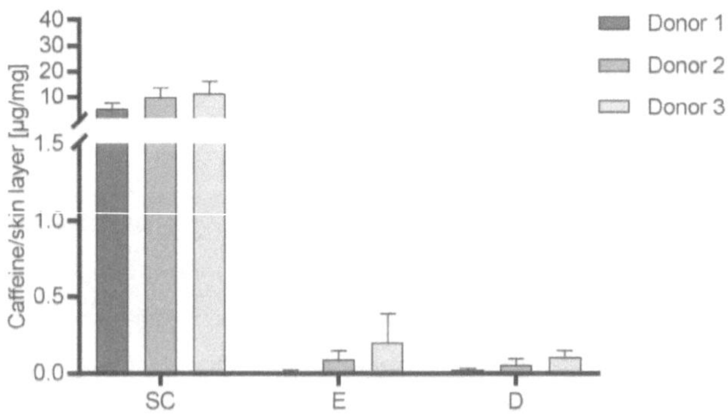

Figure 39: The mean amount of caffeine per skin layer within the SC, E and D of 500 µm porcine split-skin for three different donors was determined. A finite dose of 10 µL/cm² of 0.7 % active solved in a PGOA formulation was topically applied for 4 h at 32°C. Each bar shows the amount of absolute mass per skin layer (SC, E and D) and represents the penetration for six replicates (n = 6) from a single donor. Values represent mean ± SD and a total recovery of the API > 70 % of the application amount.

Figure 39 shows the mean amount of caffeine which penetrated in the different skin layers for a finite topical application of a PGOA formulation for three different donors. For each donor the highest amount of caffeine was found in the SC and the E shows a higher amount compared to the D, per mass of the skin layer. All three donors show an intra-donor equilibrium and a donor-specific trend for the different skin layers (Table 5). The concentration of active ingredient that penetrated into the E and D corelated with the amount of active ingredient in the SC for a specific donor. The trend if and how much caffeine penetrated each skin layer shows a similar correlation for each donor.

Table 5: Amount of caffeine per skin layer weight for three donors inside the SC, E, D, E/D and SC/E/D.

Caffeine [µg/mg]	Donor 1	Donor 2	Donor 3
SC	5.14	9.85	11.36
E	0.01	0.09	0.2
D	0.02	0.05	0.11
E/D	0.03	0.14	0.31
SC/E/D	5.17	9.99	11.67

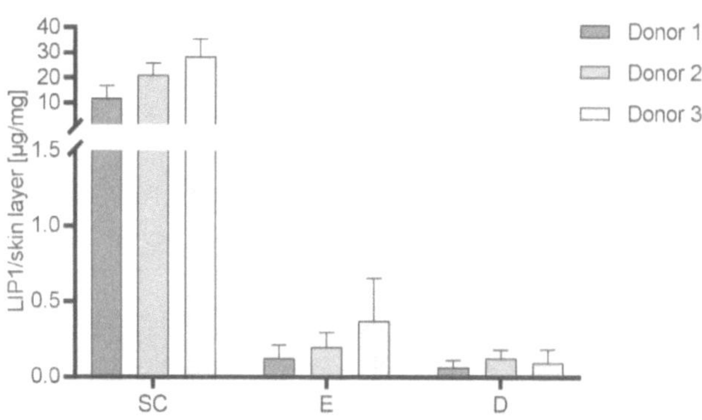

Figure 40: The mean amount of LIP1 per skin layer within the SC, E and D of 500 µm porcine split-skin skin for three different donors was determined. A finite dose of 10 µL/cm² of 0.7 % active solved in a PGOA formulation was topically applied for 4 h at 32°C. Each bar shows the amount of absolute mass per skin layer (SC, E and D) and represents the penetration for six replicates (n = 6) from a single donor. Values represent mean ± SD and a total recovery of the API > 70 % of the application amount.

Figure 40 shows the mean amount of LIP1 which penetrated in the different skin layers using a finite topical application of a PGOA formulation for three different donors. For each donor the highest amount of LIP1 was determined in the SC and the E shows a higher amount than the D, per mass of the skin layer. All three donors show an intra-donor equilibrium for the different layers (Table 6). If in the SC a lower amount of active was determined in the E also a lower amount was found in the D of one donor, except for the D of donor three shows a lower amount of LIP1. The trend if and how much caffeine penetrated each skin layer shows a similar correlation for each donor.

Table 6: Amount of LIP1 per skin layer weight for three donors inside the SC, E, D, E/D and SC/E/D.

LIP1 [µg/mg]	Donor 1	Donor 2	Donor 3
SC	11.84	21.12	28.4
E	0.12	0.2	0.37
D	0.07	0.13	0.11
E/D	0.19	0.33	0.48
SC/E/D	12.03	21.45	28.88

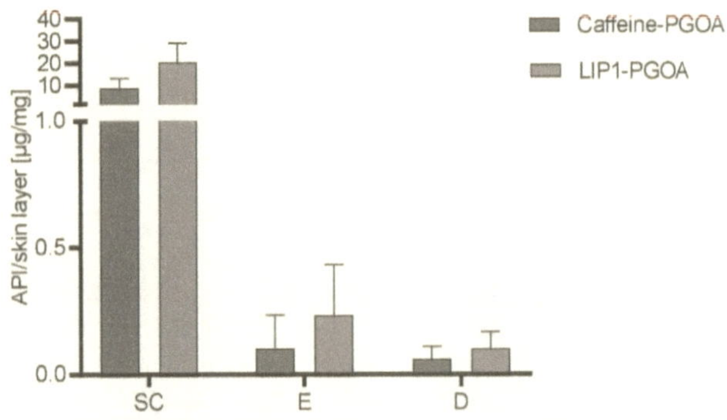

Figure 41: The mean amount of caffeine (blue) and LIP1 (red) per skin layer within the SC, E and D of 500 µm porcine split-skin skin for three different donors was determined. A finite dose of 10 µL/cm² of 0.7 % active solved in a PGOA formulation was topically applied for 4 h at 32°C. Each bar shows the amount of absolute mass per skin layer (SC, E and D) and represents the penetration for six replicates (n = 6) from a single donor. Values represent mean ± SD and a total recovery of the API > 70 % of the application amount.

Figure 41 shows the mean of caffeine and LIP1 amount within the different layers of skin for a finite topical application of a PGOA formulation for three donors. For both actives the SC> E> D amount of caffeine and LIP1. Caffeine shows a 87-fold higher amount in the SC than the E and the E a 1.7-fold higher amount than the D. LIP1 shows a 89-fold higher amount in the SC than the E and the E a 2.3-fold higher amount than the D. Inside each layer a significant higher amount of LIP1 was determined compared to caffeine. The SC shows a 2.4-fold, the E a 2.3-fold and the D a 1.7-fold higher amount for LIP1 than caffeine. The SD of the E and D layer was higher than for the SC.

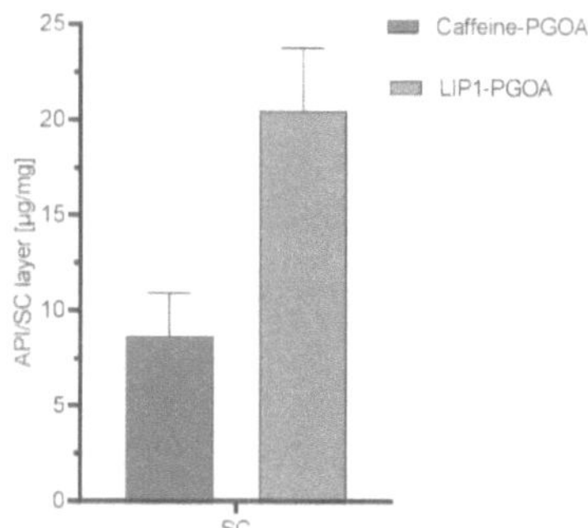

Figure 42: The mean amount of caffeine (blue) and LIP1 (red) inside the SC of 500 μm porcine split-skin for three different donors was determined. A finite dose of 10 μL/cm² of 0.7 % active solved in a PGOA formulation was topically applied for 4 h at 32°C. Each bar shows the amount of absolute mass per skin layer (SC, E and D) and represents the penetration for six replicates (n = 6) from a single donor. Values represent mean ± SD and a total recovery of the API > 70 % of the application amount.

Figure 42 shows the mean of caffeine and LIP1 amount within the SC for a finite topical application of a PGOA formulation for three donors. Inside the SC there was 8.6 μg/mg and 20.5 μg/mg per SC tissue found for caffeine and LIP1. LIP1 shows with a 2.4-fold penetration ability inside the SC a significant higher amount.

6.2.1.5 Penetration kinetics for an infinite applied formulation volume

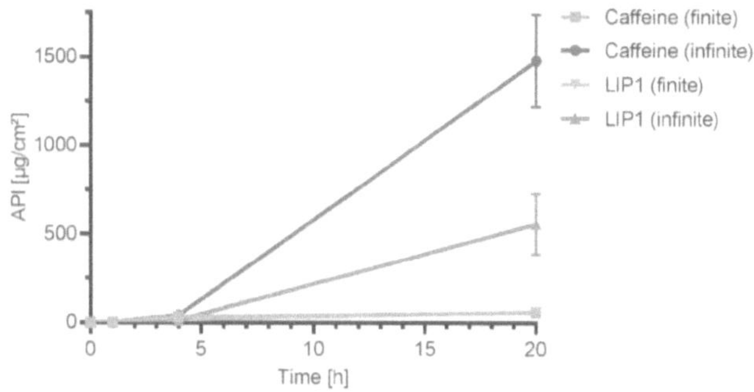

Figure 43: The kinetic penetration of caffeine (blue) and LIP1 (red) through 500 μm porcine split-skin for 1 h, 4 h and 20 h at 32°C was determined. Every time point represents the penetration flux for six replicates (n = 6) from a single donor for a topical application of finite (10 μL/cm²) or infinite (786 μL/cm²) of 0.7 % active in a PGOA formulation. Values represent mean ± SD and a total recovery of the API > 70 % of the application amount.

Figure 43 shows the kinetic penetration flux of caffeine and LIP1 for a finite and infinite topical application of a PGOA formulation over time through porcine skin. The penetration of the active into the skin was increased by using an infinite amount of formulation. Caffeine penetrated for the first 4 h the skin in a higher amount than LIP1 for an infinite and finite application. After 4 h a higher amount of LIP1 penetrated through the skin for an infinite application than for the finite caffeine or LIP1 application. Caffeine penetrated the skin with a flux of 1480 µg/cm² for an infinite and 58 µg/cm² for a finite application after 20 h. LIP1 penetrated the skin with a flux of 558 µg/cm² for an infinite and 60 µg/cm² for a finite application after 20 h. After 20 h the infinite applied caffeine penetrated 2.7-fold higher than LIP1 and LIP1 1.1-fold higher than caffeine for a finite applied dose.

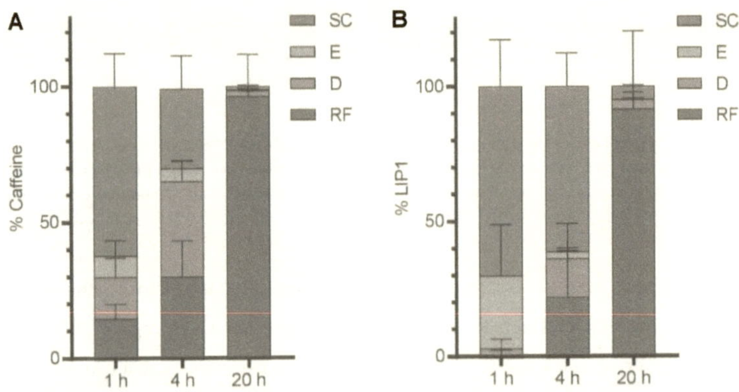

Figure 44: The mean distribution of caffeine (A) and LIP1 (B) within the SC, E and D of 500 µm porcine split-skin after 1 h, 4 h and 20 h at 32 °C was determined. An infinite dose of 786 µL/cm² of 0.7 % active solved in a PGOA formulation was topically applied. Each bar shows the percentage of determined active in each layer (Lateral, SW, SC, E, D and RF) and represents the penetration for six replicates (n = 6) from a single donor. Values represent mean ± SD and a total recovery of the API > 70 % of the application amount.

Figure 44 shows the penetration of caffeine and LIP1 for an infinite topical application of a PGOA formulation. The figure shows an increase of caffeine and LIP1 in the RF over time with a higher amount and faster penetration for caffeine. After 20 h 96 % of determined caffeine and 92 % of determined LIP1 is located in the RF. Over time the amount of determined caffeine and LIP1 which penetrated the skin decreases inside the SC and E. For the 1 h and 4 h penetration experiment the determined amount of LIP1 is with 70 % and 61 % higher than the 63 % and especially the 31 % of caffeine. The determined amount of caffeine inside the D is with 36 % after 4 h higher than 15 % and 2 % after 1 h and 20 h. For LIP1 14 % was in the dermis after 4 h and after 1 h and 20 h 3 % and 5 % were determined.

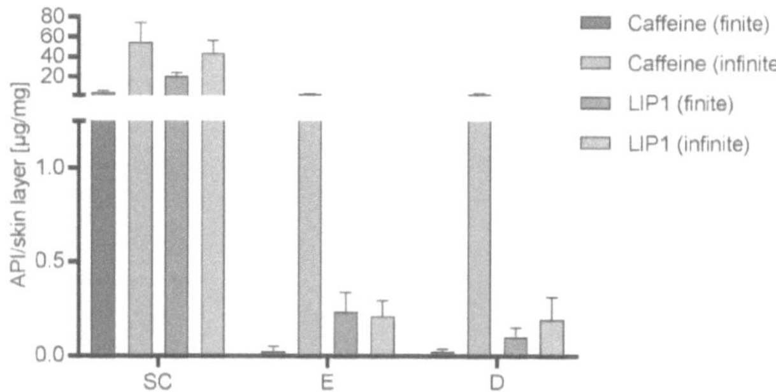

Figure 45: The mean amount of caffeine (blue) and LIP1 (red) per skin layer within the SC, E and D of 500 µm porcine split-skin after 4 h at 32°C was determined. An infinite (lighter) dose of 786 µL/cm² and finite (darker) dose of 10 µL/cm² of 0.7 % active solved in a PGOA formulation were topically applied. Each bar shows the amount of mass per skin layer (SC, E and D) and represents the penetration for six replicates (n = 6) from a single donor. Values represent mean ± SD and a total recovery of the API > 70 % of the application amount.

Figure 45 shows the mean amount of caffeine and LIP1 which penetrated in the different skin layers for a finite and infinite topical application of a PGOA formulation for 4 h. For both actives and all skin layers more active penetrated through the skin when an infinite amount was applied except for the LIP1 penetration for a finite amount in the E. Caffeine penetrated into the SC 13-fold, the E 100-fold and the D 118-fold significant more and LIP1 penetrated into the SC 2.1-fold and the D 2-fold significant more and the E 0.9-fold less.

6.2.2 Skin penetration using the MD setup

For the percutaneous skin penetration experiments *ex vivo* porcine skin was used at room temperature with the MD setup and PBS++ as receptor fluid at a flow rate of 3 µL/min. Therefore, the penetration of a finite amount (10 µL/cm²) of caffeine or LIP1 from a PGOA formulation into MD membranes implanted either superficially or deep in the skin as well as their lateral penetration were tested.

6.2.2.1 AR of API in the membrane & recovery of API in the membrane for skin saturation

For a validation of the permeation ability of caffeine and LIP1 inside the MD membrane in a PG and PGOA formulation the AR was determined. Therefore, the AR shows the theoretical available amount of active which could penetrate through the membrane. To distinguish the skin penetration ability the API containing PG and PGOA formulation was injected inside the porcine skin next to the MD membrane to test the SR ability. Due to this the maximum realistic amount of active expected by penetration through the skin into the membrane is distinguished.

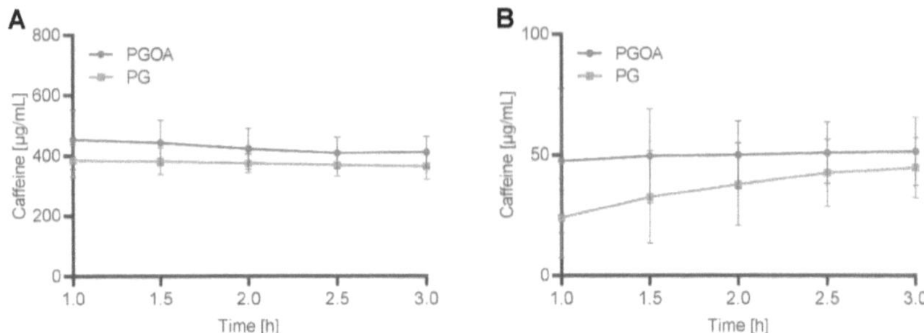

Figure 46: Absolute & skin recovery of caffeine. For the AR (A) an infinite amount (1000 µL) of a 0.7 % caffeine in a PG and PGOA formulation and the membrane were inserted in a universal tube. For the SR (B) a finite amount (125 µL) of a 0.7 % caffeine in a PG and PGOA formulation was injected inside the skin. The mean amount of caffeine over 2 h at room temperature was determined every 30 min with a flow rate of 3 µL/min. Each measurement represents the amount of caffeine penetrated per mL receptor fluid PBS++ for three replicates (n = 3). Values represent mean ± SD, compared using t-Test statistics.

Figure 46 shows the absolute and skin recovered concentration of recovered caffine applied in a PG and PGOA formulation. For the PG and PGOA formulation, the AR of caffeine over 2 h shows a linear trend and no statistical significant difference in the permeation ability. PGOA shows a caffeine permeation range of 412 to 454 µg/mL and PG a 364 to 385 µg/mL. For the PG and PGOA formulation, the SR of caffeine over 2 h shows a linear trend and no statistical significant difference in the penetration ability. PGOA shows a caffeine penetration range of 48 to 51 µg/mL and PG a 24 to 45 µg/mL. The ability of caffeine penetration is for PGOA in a AR 8.1-fold higher than for SR and PG is 8.1-fold higher.

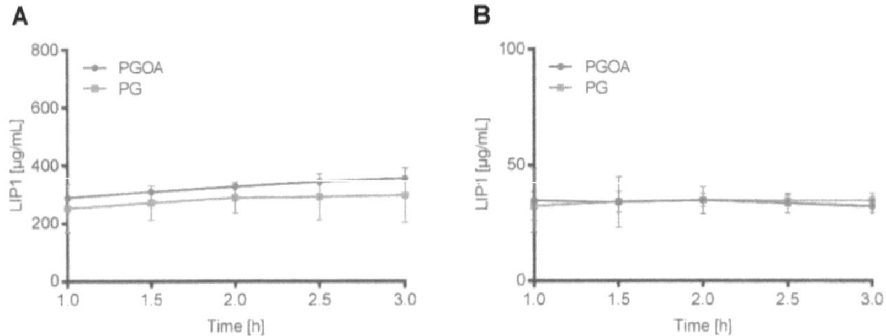

Figure 47: Absolute & Skin recovery of LIP1. For the AR (A) an infinite amount (1000 µL) of a 0.7 % LIP1 in a PG and PGOA formulation and the membrane were inserted in a universal tube. For the SR (B) a finite amount (125 µL) of a 0.7 % LIP1 in a PG and PGOA formulation was injected inside the skin. The mean amount of LIP1 over 2 h at room temperature was determined every 30 min with a flow rate of 3 µL/min. Each measurement represents the amount of LIP1 penetrated per mL receptor fluid PBS++ for three replicates (n = 3). Values represent mean ± SD, compared using t-Test statistics.

Figure 47 shows the absolute & skin recovered concentration for LIP1 applied in a PG and PGOA formulation. For the PG and PGOA formulation the AR of LIP1 over 2 h shows a linear trend and no statistical significant difference in the permeation ability. PGOA shows a LIP1 permeation range of 288 to 355 µg/mL and PG a 253 to 297 µg/mL. The SR of caffeine shows for the PG and PGOA formulation over 2 h a linear trend and no statistical significant difference in the penetration ability. PGOA shows a caffeine penetration range of 32 to 34 µg/mL and PG a 32 to 35 µg/mL. The ability of LIP1 penetration is for PGOA in a AR 11.1-fold higher than for SR and PG is 8.5-fold higher.

6.2.2.2 Different penetration depth inside the skin

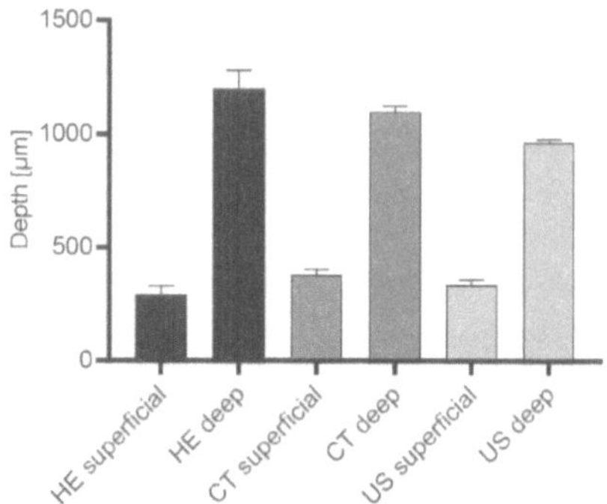

Figure 48: MD membrane depth inside skin. The depth of a superficial and deep implanted MD membrane inside porcine skin was determined using HE, CT and US. Values represent mean ± SD, compared using one-way ANOVA statistics.

Figure 48 shows the depth of the superficial and deep implanted MD membrane inside porcine skin. For the determination of the superficial implanted MD membrane HE, CT and US show a depth of 294 µm, 380 µm, 338 µm and for the deep implanted a depth of 1199 µm, 1098 µm, and 966 µm. Therefore, the superficial depth determination varies 23 % and 20 % for deeply implanted membranes.

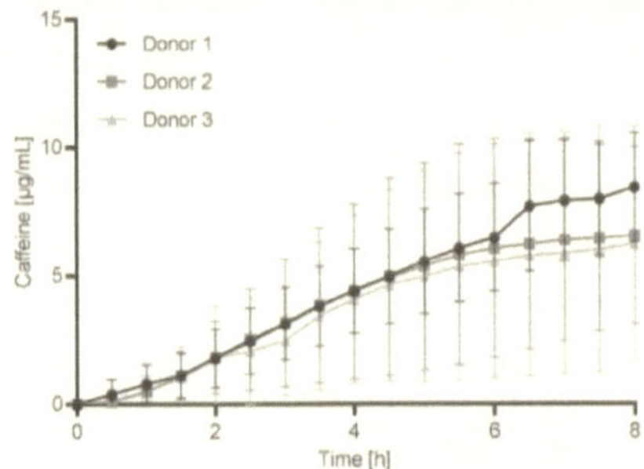

Figure 49: Caffeine penetration into porcine skin using a superficial implanted MD. A finite amount (10 µL/cm²) of a 0.7 % caffeine PGOA formulation was topically applied with a window color fence for three different donors sampled with a superficial implanted MD membrane. The mean amount of caffeine over 8 h at room temperature was determined every 30 min with a flow rate of 3 µL/min. Each measurement represents the amount of caffeine penetrated per mL receptor fluid PBS++ for six replicates (n = 6) from a single donor. Values represent mean ± SD, compared using one-way ANOVA statistics.

Figure 49 shows the amount of caffeine which penetrated inside the porcine skin and was determined after sampling from the superficial MD membrane for a finite amount of PGOA formulation for three different donors. All three donors show no statistically significant differences for the penetration of caffeine inside the skin. A linear increase between 0-6 h reaching a plateau after 8 h is shown and 6.2-8.4 µg/mL and a difference of 2.2 µg/mL caffeine was detected in the membrane after 8 h. The half of the maximum is reached after 3.8 h with 3.8-4.1 µg/mL caffeine and a difference of 0.3 µg/mL. The range of the half maximum is 14 % of the maximum variance. The SD of donor 1< donor 2< donor 3.

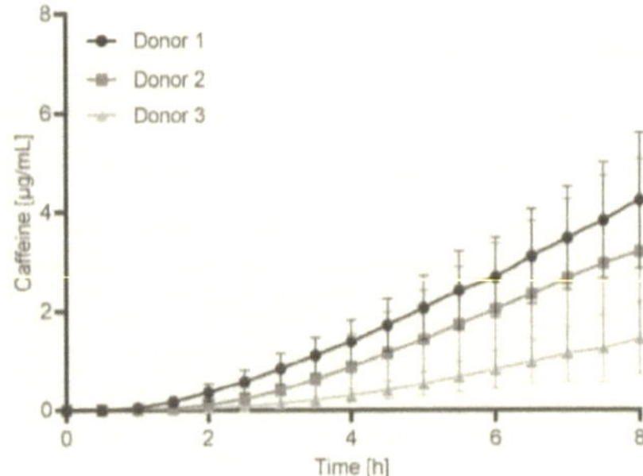

Figure 50: Caffeine penetration into porcine skin using a deep implanted MD. A finite amount (10 µL/cm²) of a 0.7 % caffeine PGOA formulation was topically applied with a window color fence for three different donors sampled with a deep implanted MD membrane. The mean amount of caffeine over 8 h at room temperature was determined every 30 min with a flow rate of 3 µL/min. Each measurement represents the amount of caffeine penetrated per mL receptor fluid PBS++ for six replicates (n = 6) from a single donor. Values represent mean ± SD, compared using one-way ANOVA statistics.

Figure 50 shows the amount of caffeine which penetrated inside the porcine skin and was determined after sampling from the deep implanted MD membrane for a finite PGOA formulation for three different donors. All three donors show no statistically significant differences for the penetration of caffeine inside the skin. A linear increase between 2-8 h is shown and 1.4-4.2 µg/cm² caffeine was detected in the membrane after 8 h which is a difference of 2.8 µg/mL. The half of the maximum is reached after 5 h with 0.5-2 µg/mL caffeine which is a difference of 1.5 µg/mL. The difference of the half maximum is 54 % of the maximum variance. The SD of donor 2< donor 1< donor 3.

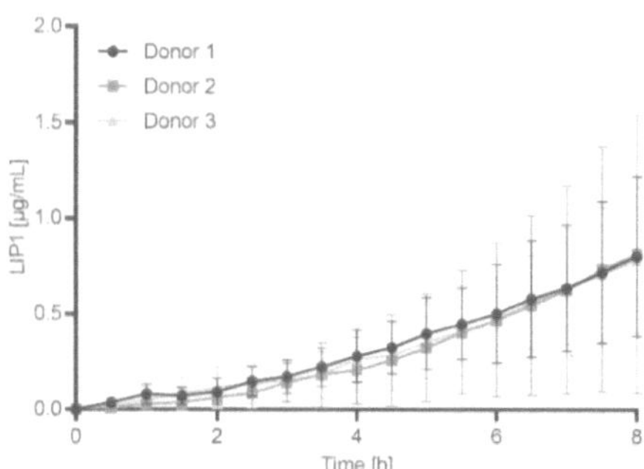

Figure 51: LIP1 penetration into porcine skin using a superficial implanted MD. A finite amount (10 µL/cm²) of a 0.7 % LIP1 PGOA formulation was topically applied with a window color fence for three different donors sampled with a superficial implanted MD membrane. The mean amount of LIP1 over 8 h at room temperature was determined every 30 min with a flow rate of 3 µL/min. Each measurement represents the amount of LIP1 penetrated per mL receptor fluid PBS++ for six replicates (n = 6) from a single donor. Values represent mean ± SD, compared using one-way ANOVA statistics.

Figure 51 shows the amount of LIP1 which penetrated inside the porcine skin and was determined after sampling from the superficial implanted MD membrane for a finite PGOA formulation for three different donors. All three donors show no statistically significant differences for the penetration of LIP1 inside the skin. A linear increase between 2-8 h is seen and 0.78-0.81 µg/cm² LIP1 was detected in the membrane after 8 h which is a difference of 0.03 µg/mL. The half of the maximum is reached after 5.5 h with 0.41-0.45 µg/mL LIP1 which is a difference of 0.04 µg/mL. The difference of the half maximum is 33 % higher than the maximum variance. The SD of donor 3< donor 1< donor 2.

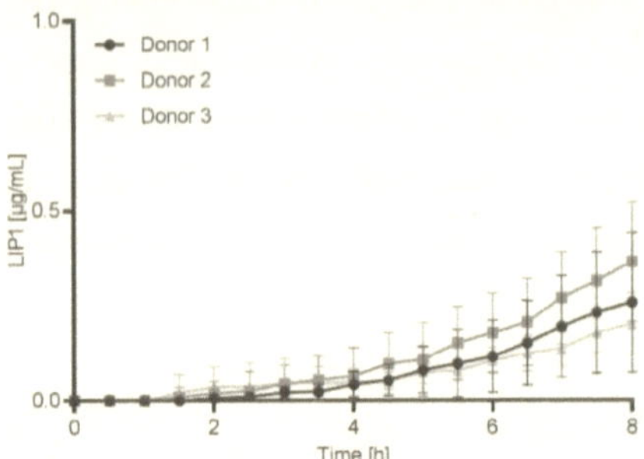

Figure 52: LIP1 penetration into porcine skin using a deep implanted MD. A finite amount (10 µL/cm²) of a 0.7 % LIP1 PGOA formulation was topically applied with a window color fence for three different donors sampled with a deep implanted MD membrane. The mean amount of LIP1 over 8 h at room temperature was determined every 30 min with a flow rate of 3 µL/min. Each measurement represents the amount of LIP1 penetrated per mL receptor fluid PBS++ for six replicates (n = 6) from a single donor. Values represent mean ± SD, compared using one-way ANOVA statistics.

Figure 52 shows the amount of LIP1 which penetrated inside the porcine skin and was determined after sampling from the superficial implanted MD membrane for a finite PGOA formulation for three different donors. All three donors show no statistically significant differences for the penetration of LIP1. The penetration of LIP1 increased over time and between 0.2-0.36 µg/cm² penetrated the skin in a deep MD membrane after 8 h which is a difference of 0.16 µg/mL. The half of the maximum is reached after 6 h with 0.1-0.18 µg/mL caffeine which is a difference of 0.08 µg/mL. The difference of the half maximum is 50 % of the maximum variance. The SD of donor 3< donor 2< donor 1.

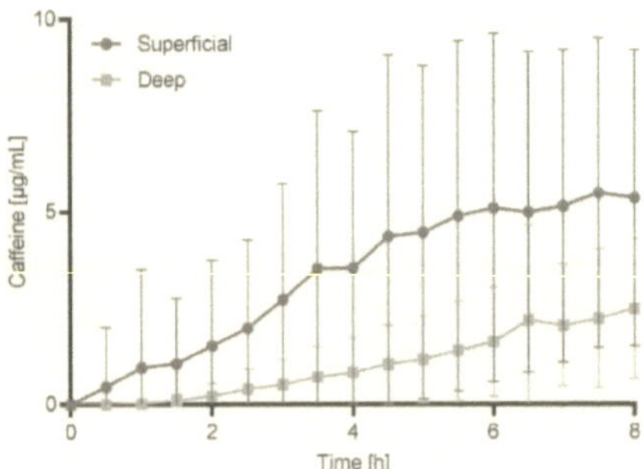

Figure 53: Caffeine penetration into porcine skin using a superficial and deep implanted MD. A finite amount (10µL/cm²) of a 0.7 % caffeine PGOA formulation was topically applied with a window color fence for three different donors sampled with a superficial (dark) and deep (light) implanted MD membrane. The mean amount of caffeine over 8 h at room temperature was determined every 30 min with a flow rate of 3 µL/min. Each measurement represents the mean amount of caffeine penetrated per mL receptor fluid PBS++ for six replicates (n = 6) for three donors. Values represent mean ± SD, compared using one-way ANOVA statistics.

Figure 53 shows the mean amount of caffeine which penetrated inside the porcine skin and was determined after sampling from the superficial and deep MD membrane for a finite PGOA formulation for three donors. A higher amount of caffeine penetrated inside the superficial membrane compared to the deep implanted membrane. After 8 h the increasing penetration of caffeine shows a plateau for the superficial membrane and an increasing penetration for the deep implanted membrane. After 8 h 5.4 µg/mL caffeine penetrated inside the superficial and 2.5 µg/mL in the deep implanted membrane with a 2.2-fold higher amount. The half of the maximum is reached after 3.1 h whereby the superficial membrane shows 4.4-fold higher amount of caffeine.

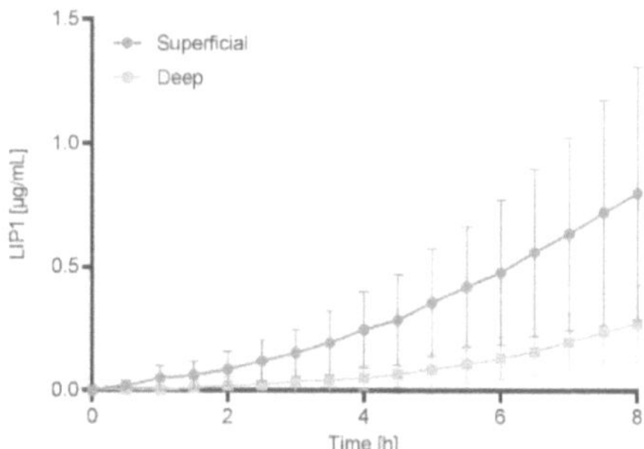

Figure 54: LIP1 penetration into porcine skin using a superficial and deep implanted MD. A finite amount (10 µL/cm²) of a 0.7 % LIP1 PGOA formulation was topically applied with a window color fence for three different donors sampled with a superficial (dark) and deep (light) implanted MD membrane. The mean amount of LIP1 over 8 h at room temperature was determined every 30 min with a flow rate of 3 µL/min. Each measurement represents the mean amount of LIP1 penetrated per mL receptor fluid PBS++ for six replicates (n = 6) for three donors. Values represent mean ± SD, compared using one-way ANOVA statistics.

Figure 54 shows the mean amount of LIP1 which penetrated inside porcine skin for three donors and was determined after sampling from the superficial and deep MD membrane for a finite PGOA formulation. A higher amount of LIP1 penetrated inside the superficial membrane compared to the deep implanted membrane. The penetration of LIP1 increased for 8 h and with 0.8 µg/mL after 8 h for the superficial and 0.27 µg/mL for the deep membrane, a 3-fold higher amount penetrated in the superficial membrane. The half of the maximum is reached after 5.4 h whereby the superficial membrane shows with 0.4 µg/mL and the deep implanted membrane with 0.07 µg/mL a 5.7-fold higher amount of LIP1.

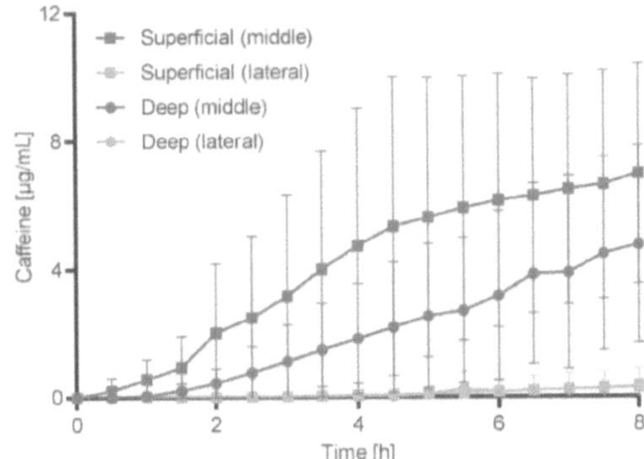

Figure 55: Lateral caffeine penetration into porcine skin using a superficial and deep implanted MD. A finite amount (10µL/cm²) of a 0.7 % caffeine PGOA formulation was topically applied with a window color fence for two different donors sampled with a superficial and two superficial lateral implanted MD membranes and deep and two deep lateral implanted MD membranes. The mean amount of caffeine over 8 h at room temperature was determined every 30 min with a flow rate of 3 µL/min. Each measurement represents the mean amount of caffeine penetrated per mL receptor fluid PBS++ for six replicates (n = 6) from two donors. Values represent mean ± SD, compared using one-way ANOVA statistics.

Figure 55 shows the mean amount of caffeine which penetrated inside and laterally the porcine skin. The amount of caffeine was determined after sampling from the superficial and deep implanted MD membrane for vertical and lateral penetration for a finite PGOA formulation. Increasing amounts of caffeine penetrated inside the superficial and deep membranes in the middle of the application area. Therefore, the penetration of caffeine in the superficial membrane increases up to a plateau and for the deep membrane the amount increases linear with a 1.5-fold higher amount of caffeine inside the superficial membrane after 8 h. The half of the maximum is reached after 3.2 h whereby the superficial membrane shows with 3.5 µg/mL and the deep implanted membrane with a 1.3 µg/mL a 2.7-fold higher amount of caffeine. The relative lateral penetration of caffeine was after 8 h 4.3 % for the superficial depth and 7 % for the deep implanted membrane.

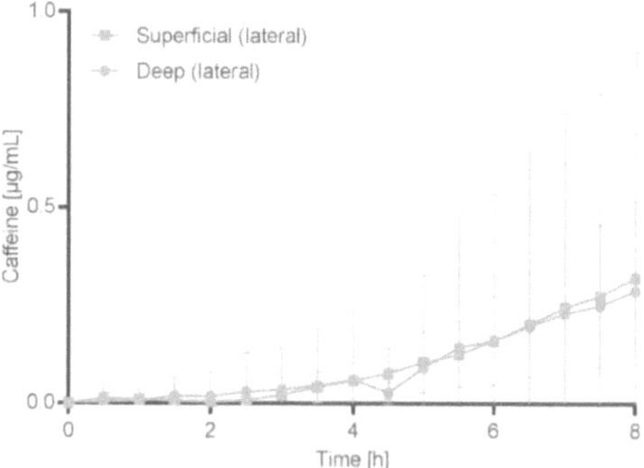

Figure 56: Lateral caffeine penetration into porcine skin using a deep implanted MD. A finite amount (10 µL/cm²) of a 0.7 % caffeine PGOA formulation was topically applied with a window color fence for two different donors sampled with two superficial lateral and two deep lateral implanted MD membranes. The mean amount of caffeine over 8 h at room temperature was determined every 30 min with a flow rate of 3 µL/min. Each measurement represents the mean amount of caffeine penetrated lateral per mL receptor fluid PBS++ for six replicates (n = 6) from two donors. Values represent mean ± SD, compared using one-way ANOVA statistics.

Figure 56 shows the mean amount of caffeine which penetrated laterally the porcine skin inside the superficial and deep MD membrane for a finite PGOA formulation. Therefore, the penetration of caffeine increased over time and after 8 h 0.32 µg/mL for the superficial and 0.29 µg/mL for the deep membrane show a similar amount penetrated laterally the skin. The half of the maximum amount which penetrated the skin was reached after 5.8 h.

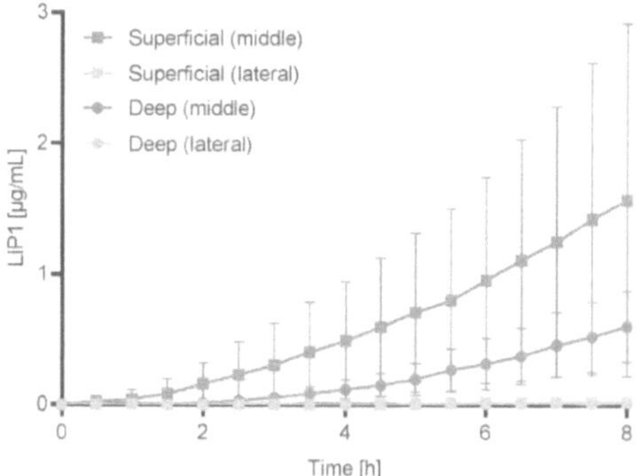

Figure 57: Lateral LIP1 penetration into porcine skin using a superficial and deep implanted MD. A finite amount (10 µL/cm²) of a 0.7 % LIP1 PGOA formulation was topically applied with a window color fence for two different donors sampled with a superficial and two superficial lateral implanted MD membranes and deep and two deep lateral implanted MD membranes. The mean amount of LIP1 over 8 h at room temperature was determined every 30 min with a flow rate of 3 µL/min. Each measurement represents the mean amount of LIP1 penetrated per mL receptor fluid PBS++ for six replicates (n = 6) from two donors. Values represent mean ± SD, compared using one-way ANOVA statistics.

Figure 57 shows the mean amount of LIP1 which penetrated inside and laterally the porcine skin. The amount of LIP1 was determined in the superficial and deep implanted MD membrane for vertical and lateral penetration for a finite PGOA formulation. Increasing amounts of LIP1 penetrated inside the superficial and deep membranes in the middle of the application area. Increasing amounts of LIP1 penetrated inside the superficial and deep vertically in the middle of the application area, the amount in the superficial membrane increases with a 2.7-fold higher amount after 8 h. The half of the maximum is reached after 5.4 h where the superficial membrane shows with 0.8 µg/mL and the deep implanted membrane with a 0.25 µg/mL a 2.7-fold higher amount of LIP1. The relative lateral penetration of LIP1 was after 8 h 1.3 % for the superficial depth and 3.3 % for the deep implanted membrane.

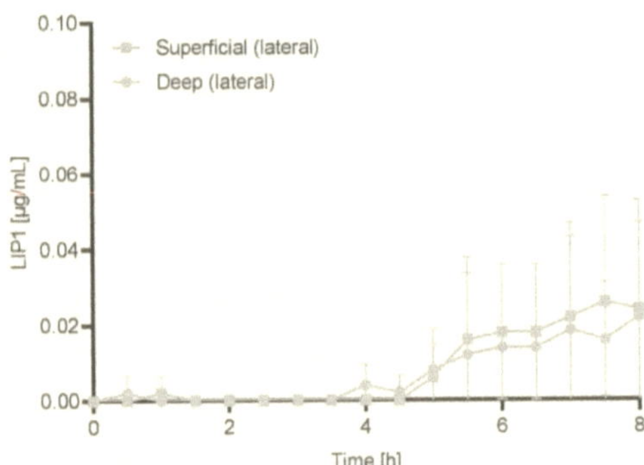

Figure 58: Lateral LIP1 penetration into porcine skin using a deep implanted MD. A finite amount (10µL/cm²) of a 0.7 % LIP1 PGOA formulation was topically applied with a window color fence for two different donors sampled with two superficial lateral and two deep lateral implanted MD membranes. The mean amount of LIP1 over 8 h at room temperature was determined every 30 min with a flow rate of 3 µL/min. Each measurement represents the mean amount of LIP1 penetrated lateral per mL receptor fluid PBS++ for six replicates (n = 6) from two donors. Values represent mean ± SD, compared using one-way ANOVA statistics.

Figure 58 shows the mean amount of LIP1 which penetrated laterally the porcine skin inside the superficial and deep MD membrane for a finite PGOA formulation. Therefore, the penetration of LIP1 increased over time and after 8 h 0.22 µg/mL for the superficial and for the deep membrane show a similar amount penetrated laterally the skin. The half of the maximum amount which penetrated the skin was reached after 5.4 h.

6.2.3 Static lateral penetration

To test the lateral percutaneous skin penetration *ex vivo* porcine skin was used at room temperature. The penetration features into the E and D of caffeine and LIP1 for a finite amount (10 µL/cm²) of PGOA formulation were tested and lateral penetration in E and D were compared to the absolute amount in these layers.

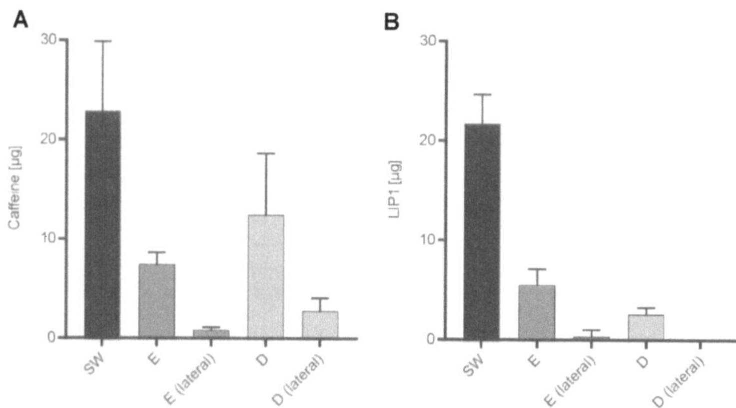

Figure 59: Lateral and vertical caffeine (A) and LIP1 (B) penetration into different porcine skin layers using a static lateral penetration setup. A finite amount (10 µL/cm²) of a 0.7 % active in PGOA formulation was topically applied with a window color fence for three different donors. The mean amount of caffeine and LIP1 over 4 h at room temperature which penetrated vertical and lateral in the E and D were determined. Each bar represents the mean amount of active penetrated the skin for two replicates (n = 2) from three donors. Values represent mean ± SD and a total recovery of the API > 70 % of the application amount.

Figure 59 shows the vertical and lateral penetration of caffeine and LIP1 inside the E and D. The SW for caffeine and LIP1 results in 22 µg. For caffeine 7.5 µg penetrated vertically in the E and 12.5 µg in the D, the lateral penetration of caffeine shows 0.85 µg in the E and 2.8 µg in the D which is 11.3 % and 22.4 %. LIP1 penetrated with 5.6 µg vertical in the E and 2.7 µg in the D, the lateral penetration of LIP1 shows 0.35 µg in the E and 0 µg in the D which is 6.3 % and 0 %. The lateral penetration for caffeine is inside the E 2.4-fold higher compared to LIP1.

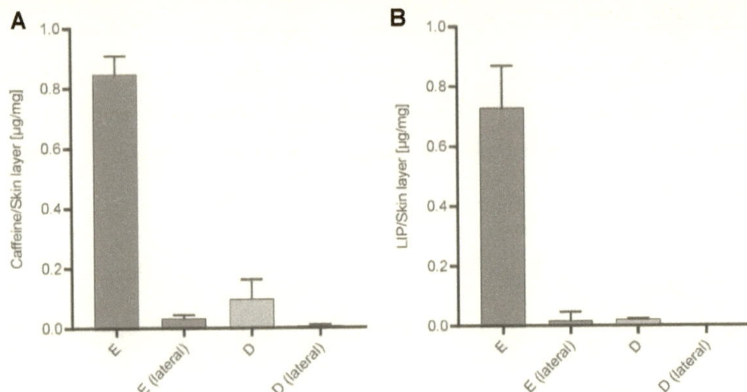

Figure 60: Lateral and vertical caffeine (A) and LIP1 (B) penetration into porcine skin per skin layer weight using a static lateral penetration setup. Lateral and vertical caffeine and LIP1 penetration into different porcine skin layers. A finite amount (10 µL/cm²) of a 0.7 % active in PGOA formulation was topically applied with a window color fence for three different donors. The mean amount of caffeine and LIP1 over 4 h at room temperature which penetrated vertical and lateral in the E and D were determined and are shown per tissue layer weight. Each bar represents the mean amount of active penetrated the skin for two replicates (n = 2) from three donors. Values represent mean ± SD and a total recovery of the API > 70 % of the application amount.

Figure 60 shows the vertical and lateral penetration of caffeine and LIP1 per E and D weight. For caffeine 0.85 µg/mg penetrated vertical in the E and 0.1 µg/mg in the D, the lateral penetration of caffeine shows 0.034 µg/mg in the E and 0.007 µg/mg in the D which is 4 % and 7 % per skin layer weight. LIP1 penetrated with 0.73 µg/mg vertically in the E and 0.02 µg/mg in the D, the lateral penetration of LIP1 shows 0.016 µg/mg in the E and 0 µg/mg in the D which is 2.2 % and 0 %. The lateral penetration for caffeine is inside the E 2.1-fold higher comparted to LIP1.

7 Discussion

Several studies have been conducted to study the dermal and transdermal delivery of drugs through the skin *in silico, in vitro, ex vivo,* and *in vivo* [38]. *In vitro* and *ex vivo* percutaneous penetration studies commonly use artificial membranes, porcine skin, and human split-skin as models to verify the penetration of topical applied APIs. In addition to external factors, the targeted penetration of active ingredients is influenced by the physico-chemical characteristics of the API as well as the formulation.

Influences on the skin penetration of an active ingredient

The formulation, formulation volume, active ingredient concentration, temperature, and receptor fluid are important for skin penetration. These parameters have been tested for their influence on the penetration rate of the model drug caffeine.

Formulations which are the vehicle of the API, are crucial for their penetration ability; they need to dissolve the active ingredient and deliver it to its target-site [69]. Formulations aimed at cosmetic use contain multiple ingredients; vehicles for transdermal delivery testing are, in contrast, not as complex. Water, PG, OA, and their mixtures are commonly used vehicles in penetration studies [142]. OA was shown to increase the penetration of both, hydrophilic and hydrophobic drugs [152]. Most formulations are water based, due to this, an understanding of the penetration ability of an aqueous vehicle is essential. PG is a component of multiple formulations due to its good solubility characteristics and multifunctional use in formulation development; it is well studied and often used in combination with fatty acids and penetration enhancers such as OA [143]. For the penetration studies the artificial Strat-M® membrane, porcine and human split-skin were used and the penetration characteristics of an infinite amount of caffeine in water, PG, or PGOA were tested. PG as a vehicle enhanced the penetration ability of the active ingredient more than water as a vehicle (Figure 26). When combined with OA, the PGOA formulation showed the highest penetration rate for caffeine. For all human skin surrogates, a similar penetration enhancing effect for the water, PG, and PGOA formulations were obtained. PG is known to enhance skin penetration by interacting with the polar headgroups of the SC lipids, whereas OA enhances the penetration by interacting with the SC lipid alkyl domain, and these interactions increases the penetration ability of the API [143]. Designing a formulation exclusively based on the physico-chemical solubility properties of the API does not optimize the delivery of the API in a cosmetic formulation, and the efficacy of a formulation can be improved by a selection of multiple ingredients. The formulation allows to optimize the delivery of active ingredients into the skin but it needs to be designed individually for each API. The API diffuses within the formulation to the formulation-skin interface and permeates through the lipophilic SC. It diffuses through the hydrophilic

E, penetrates into the hydrophilic D, and culminates in delivery to the systemic circulation, which leads to systemic absorption. [51]

Changing the application volume with a constant API concentration increases the penetration rate for an increasing volume (Figure 27). The penetration rate of caffeine into porcine skin increased linearly with higher application amounts. With the Strat-M® membrane, penetration first increased linearly with API volumes less than 100 µL/cm² and then plateaued as the amount of the API administered increases. In contrast, the porcine skin showed a linear increase in penetration rate with higher application amounts of caffeine (Figure 27). An increasing penetration rate for higher application volumes has multiple reasons. Increasing the volume leads to higher skin occlusion, which leads to inhibition of skin evaporation and swelling of the SC layer, as well as an increase in the skin temperature [3, 21, 147]. This has a direct influence on penetration of the active and deposition, affects the degree of hydration of the SC, and enhances penetration via occlusion [153, 154]. The concentration gradient, which is the driving force of penetration rate, remains stable long term and increases overall penetration [38]. Increasing the application volumes leads to an increased hydrostatic pressure, which increases the penetration rate [145].

Changing the concentration of topically applied API leads to a change in the concentration gradient between the skin surface, inner skin, and the receptor chamber. The concentration gradient drives diffusion of active ingredients through the skin [53]. This passive process is described by Fick´s first law (Equation 1) and increases with a higher concentration gradient. Figure 28 shows the higher caffeine penetration through the Strat-M® membrane and the porcine split-skin with the increase in the active concentration inside the formulation. Increasing the concentration of caffeine in the formulation, leads to an increase in the caffeine penetration flux [155]. Depending on the molecular properties of the API, an up to 300-fold increase in penetration was noted along with an increase in uptake. Depending on the degree of saturation of the formulation, different skin layers such as the SC show an 18-fold increase in API concentration within those layers [156, 157].

Changing the temperature of the percutaneous penetration setup changes the skin uniformity and especially the fluidity of the lipids found within the skin [8]. For an infinite amount of caffeine applied topically in a PGOA formulation, the penetration rate through the Strat-M® membrane and porcine split-skin increased as a result of a higher temperature in the setup (Figure 29). Therefore, a linear correlation between the increased penetration rate and temperature is observed for both skin surrogates. At room temperature, the main parts of the SC lipids are solid, and with increasing temperatures they become more fluid, which leads to an increased penetration rate [8]. Trabaris et al. show that the temperature increase from 25°C to 40°C, changes the skin permeability with a 3-fold higher penetration flux, in a similar magnitude increase of penetration [148].

By changing the receptor fluid composition of the FDC two main criteria are essential: (1) no alteration of the skin integrity, and (2) ensuring the chemical solubility of the API inside the receptor fluid [105]. Modified physiological receptor fluids are important to test the penetration of highly lipophilic APIs. To have universally receptor fluid for penetration testing of lipophilic and hydrophilic actives, a receptor fluid containing multiple additives like albumin, cyclodextrin or the surfactant Brij are used [105, 122, 123]. Adding albumin to the receptor fluid is recommended for hydrophilic and lipophilic APIs and does not modify the integrity of the skin [105]. Brij as a surfactant has been reported to enhance the solubility of hydrophilic and lipophilic drugs [122]. Cyclodextrin is recommended for penetration testing of multiple active ingredients and is known to increase the solubility of both hydrophilic and lipophilic drugs [123]. Those additives have been used in combination with PBS++ or water to test the penetration ability of caffeine [105, 158]. The findings show that caffeine penetrates the Strat- M® membrane and porcine split-skin using multiple different receptor fluids, which are recommended for lipophilic APIs (Figure 30 and Figure 31). The penetration rate of the hydrophilic caffeine was comparable to that of pure PBS++ when albumin, cyclodextrin, and Brij were individually added to the receptor fluid. Therefore, those receptor fluids are used for comparing the penetration rate of water insoluble lipophilic APIs with the model drug caffeine. To increase the penetration ability of APIs, especially caffeine, an optimised vehicle containing the penetration enhancer, in combination with a high application volume and high API concentration at 37°C with a suitable receptor fluid is necessary.

Suitable human skin surrogates for percutaneous penetration studies

For percutaneous penetration testing, *in vivo* and *ex vivo* human studies are the most accurate method of choice for the prediction of efficient penetration abilities. Due to ethical reasons and limited availability of human skin, artificial membranes and *ex vivo* porcine skin are suitable surrogates for *in vitro* skin studies [44]. To identify the comparability of different *in vitro* and *ex vivo* penetration studies, the artificial Strat-M® membrane, porcine ear, and human split-skin were compared and tested for their penetration ability as human skin surrogates, using caffeine in an FDC penetration setup. Therefore, skin integrity, intra- and inter-experimental variations, the impact of kinetic sampling, and the influences of penetration enhancers on the penetration rate on different human skin surrogates are important. The Strat-M® membrane is a multilayer synthetic membrane that is coated with a SC mimicking lipid layer [81]. Porcine ear and human abdomen skin is mostly used for penetration studies with a defined dermatomed thickness to achieve reproducible uniformity and predictability; the standardized thickness is 500 µm [2]. To validate the thickness of the experimental application area of the skin, the peripheral measurement sites were used to determine areas with no statistically significant differences in skin thickness (Figure 18). The integrity of the skin surrogates was determined via TEWL; an intra-surrogate uniformity of 20 ± 1, 14 ± 6, and 12.5 ± 3.5 g/m^2/h for the Strat-M$^\circ$ membrane, porcine, and human skin, respectively, were observed (Figure 19). Therefore, TEWL does not allow for a distinction in the permeability of the skin but enables the confirmation of the structural integrity of the model used. TEWL is limited in its ability to assess small changes and a decreasing TEWL for thicker *ex vivo* porcine split-skin with a slope of -0.0156 was observed (Figure 20) [159]. In Figure 21, the experimental variation is shown for all human skin surrogates and the intra-experimental variation was found to be larger than the inter-experimental variation (Figure 21 D). The intra-experimental variation was minimized by the higher number of replicates. Therefore, the minimum of four replicates recommended by the OECD was increased to a minimum of six replicates for the penetration testing [44]. The inter-experimental variations are shown for the RF as well as for all different skin layers for a lipophilic and hydrophilic molecule (Figure 37). The inter-experimental variation is described as a donor-specific difference in the skin layers and a difference in penetration ability [99]. To better understand the influences of kinetic sampling, individual experiments were compared with experimental sampling for every hour or two hours. The inter-experimental differences showed a higher impact on the penetration rate of caffeine than the experimental sampling time and no significant differences were observed for the human skin surrogates (Figure 22- 24). Similar penetration kinetics were observed for caffeine for the Strat-M$^\circ$ membrane and porcine split-skin, with a 2.8-fold difference (Figure 25). The findings were comparable to the penetration rate of caffeine using human split-skin, with a delay in time and a lower overall penetration flux, which needs to be considered when interpreting penetration ability [99]. Testing the penetration ability of different caffeine containing formulations, it was found that the penetration enhancing effectbdemonstrates the same trend for the different surrogates (Figure 26). The porcine skin showed an overall low penetration rate for caffeine, and human skin shows an even lower penetration rate [84, 85, 99, 160]. Artificial membranes and especially the Strat-M$^\circ$ membrane support penetration studies and also have the potential to help elucidate factors influencing penetration of an API [161, 162]. Porcine ear skin is a valuable human skin surrogate owing to its similar skin layer thickness and penetration, providing information on the influences on penetration of APIs, with a higher penetration rate than human skin [160].

Barrier function of the different skin layers

To study percutaneous penetration, an understanding of the barrier function of SC, E, and D as well as the diffusion and penetration ability of hydrophilic and lipophilic molecules in those layers is important. The SC as the outer most skin layer, builds a lipophilic environment with an assumed log P of 0.8 [51]. The log P describes the solubility characteristics of an API. To ensure penetration, the API needs to be just as much dissolved in the formulation but should not lose its ability to partition into the skin. After overcoming the first line of defense, which is the SC, the targeted penetration site could be the aqueous E or the D. Lipophilic compounds that overcome the SC barrier, often result in the reduction of penetration rates if the more hydrophilic E and D layer is reached [38]. The route of penetration an API undertakes after it is topical applied is influenced by absorption from the skin surface into the lipophilic SC, followed by subsequent permeation through the more aqueous viable E into the aqueous D [38]. APIs with physico-chemical characteristics of a molecular weight lower than 500 Da and a log P 1-4 are thought to possess good penetration abilities and move through the skin layers via passive diffusion [28, 61]. The API caffeine (Log P = -0.1) and LIP1 (Log P = 0.6) with an identical molecular weight of 194.19 g/mol but different lipophilic characteristics are ideal for a comprehensive comparative penetration study.

Figure 32 shows the influence of the removal of the SC and E barrier on the penetration of caffeine and LIP1 as well as the influence of different D thicknesses on penetration within porcine split-skin. The removal of the whole E increases the penetration of both, caffeine and LIP1, with a larger increase in penetration rate for caffeine. The removal of the SC had a statistically significant influence on the penetration rate of the hydrophilic caffeine, but not for the lipophilic LIP1. With varying D thickness, the penetration rate of LIP1 decreases more than that of caffeine, which leads to a linear decrease for LIP1 and a plateau for caffeine. Andrews et al. reported that the removal of the SC increases the penetration of compounds drastically and the removal of the full E increases this by another 1-2 orders of magnitude [163]. Percutaneous penetration experiments measured kinetically with a finite application of the API show that skin penetration and diffusion of active ingredients inside the different skin layers, to compare the influence of the lipophilicity for the penetration ability. The penetration kinetics of caffeine and LIP1 over 1 h, 4 h, and 20 h for a finite formulation of PG and PGOA, show over time that caffeine has a higher penetration rate through the skin tissue into the RF than LIP1, but a higher amount of active inside the SC, for all time points (Figure 34 and Figure 35). When the E and D is reached over time LIP1 shows a higher amount in those layers compared to caffeine, due to a reduction of the penetration rates the lipophilic active remains longer in those aqueous layers. Comparing in Figure 36 the amount of active inside the skin layers per skin layer weight, LIP1 shows a higher skin saturation of all different layers, when measured in a time frame where the active ingredient reached those layers. The lipophilic LIP1 shows good penetration potential into the SC, after overcoming this barrier, the penetration rate is reduced in the hydrophilic E and D environment and the active accumulates in those layers, till it is driven into the RF by the gradient over time. In Figure 38 the mean percentage per skin layer is shown and is higher in the SC, E and D for LIP1 compared to caffeine. The lipophilic character of APIs supports the penetration through the SC layer and not the overall penetration into the skin this needs to be considered by the generalization that APIs with a Log P between 1-4 show good penetration abilities [28, 61].

Increasing the topical application to an infinite amount of active in a PGOA formulation, increases the total amount of caffeine 25-fold and LIP1 9-fold inside the RF (Figure 43). Chen et al. reports the same magnitude of penetration increasement by changing the finite to an infinite volume for hydrophilic and lipophilic molecules [164]. Figure 44 shows the increasing total amount of active ingredient in all skin layers and Figure 45 the increasing amount of active inside the SC, E and D by skin layer weight. The amount of caffeine and LIP1 increases inside the SC by increasing the application volume with no statistically significant difference between the concentrations measured for caffeine and LIP1 when an infinite formulation volume was applied. In contrast the E and D results in an increase inside the tissue for caffeine and no statistically difference for LIP1 in those layers for an infinite amount of formulation. Herbig et al. reported an increasing amount of active inside the skin layers by increasing the application volume for porcine and human skin and the same magnitude of active within the E and D layer per skin tissue [165].

From the MD experiments with caffeine and LIP1, the AR and SR is higher for the hydrophilic caffeine than for the lipophilic LIP1, with a small increase in penetration for caffeine and the SR [166]. The MD setup obtains valid experimental data to compare the penetration ability of caffeine and LIP1 into the skin and more precise in different D depth [118]. To ascertain the depth of the MD membrane inside the skin, the HE staining, US and CT data are compared. The measured depth of the membrane varies for all determination methods and the average depth of a superficial implanted membrane ranges from 300-500 µm and for a deep implanted from 900-1300 µm. Each of these methods have their own advantages and disadvantages making them difficult to compare to each other. So far none of these methods has been recognized as the golden standard, and hence it would be most meaningful to select a specific method and compare MD membrane implantation depth measurements for a single method. The superficial and deep membranes are both implanted in the D layer, but actives which penetrate the deep membrane penetrate further through the D layer. Figure 53 shows the concentration of caffeine which increases over time, finally resulting in a plateau for penetration into both superficial and deep implanted membranes. Figure 54 shows the increase of concentration of LIP1 which penetrated into the superficial and deep implanted membrane over time. The half maximum amount of penetrated active was reached after 3.1 h for the hydrophilic caffeine and after 5.4 h for the lipophilic LIP1. A higher difference of penetration ability for LIP1 compared to caffeine into the superficial and deep membrane for the 8 h and half maximum amount are shown. These results reinforce the observation that the hydrophilic D environment impacts the penetration of actives and slows down the penetration rate of lipophilic molecules. It was also previously suggested by Koch et al. that the permeability of lipophilic molecules is retarded by the diffusional resistance caused by the dermal matrix. This retardation of permeation continues till the accumulation of the API results in the saturation of the tissue [167].

Intra-donor skin equilibrium

To better understand percutaneous penetration, donor-specific skin layer equilibria and donor variation was studied. The SC, E and D are tested for a donor-specific equilibria, using different donors for a standard FDC penetration setup as well as a penetration kinetic over 20 h.

The kinetic caffeine and LIP1 penetration experiments with a finite PGOA formulation show a maximum capacity of active in different layers of the skin that was specific to a given donor. Inside the E and D there is an intra-donor specific equilibrium which is shown in Figure 36 for caffeine and LIP1. Therefore, the same inter trend for an intra equilibrium for both actives are shown for six donors inside the E and D for the penetration experiments of 1 h, 4 h and 20 h. The donor-specific variance is shown in Figure 37 for three different donors and a 4 h finite topical application penetration experiment of caffeine and LIP1 in a PGOA formulation. For caffeine and LIP1 all three donors show an intra-donor trend of active concentration inside the SC, E and D layer (Table 5 and Table 6). In Figure 39 the caffeine amount in each skin layer normalised to the weight of that layer shows an intra-donor equilibrium of each donor and the active in the SC, E and D which varies inter-donor specific. The amount of LIP1 in each skin layer normalised to the weight of that layer, confirms the intra-donor equilibrium (Figure 40). This equilibrium is layer specific which is shown in Figure 45, a saturation stage and an entrapped compound reservoir in the different skin layers with a layer depended maximum capacity of a hydrophilic and lipophilic compound is observed. The constant flux of API results in a saturation concentration per skin layer with a constant in and out flow of active. Herbig et al. reports the same magnitude of active within the specific layers per tissue weight [165]. The D appears to accumulate the actives until the tissue is saturated and reaches the receptor chamber [167]. The aqueous layers incorporate less lipophilic active than the lipophilic skin layer of a hydrophilic or lipophilic active. The data supports the assumption of an intra-donor equilibrium which is donor depended, the tissue entraps the active till a saturation is reached and enables a further diffusion after. The reservoir capacity of the SC is limited and its saturation is dependent on each individual donor, application site, substance and formulation [21]. Southwell et al. reported variations of penetration in human skin within and between donors but did not observe a donor-specific equilibrium of the different skin layers [168]. A schematic illustration of an intra-donor equilibrium and a 1.5-fold higher inter-donor trend

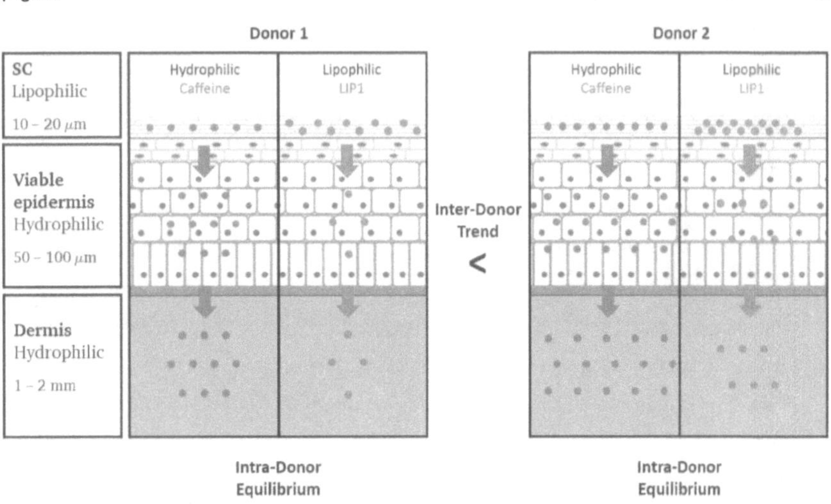

Figure 61: Schematic illustration of the intra-donor equilibrium with an inter-donor difference. Donor 2 shows a 1.5-fold higher saturation of active inside the tissue.

Effect of the lateral penetration inside the skin

The lipophilic character of an API influences the penetration through and diffusion inside the skin depending on the environmental skin layer. Lateral penetration of the hydrophilic caffeine and lipophilic LIP1 inside the aqueous E/D layer via the FDC setup and inside D layer via a MD setup and a static lateral penetration setup was investigated. For the FDC setup the lateral movement is higher over time for caffeine than for LIP1, for a finite topical application in a PG and PGOA formulation (Figure 34 and Figure 35). In Figure 38 the mean amount of active shows the lateral penetration inside the E/D layer for three different donors, with a statistically significant difference between caffeine and LIP1 and a 1.6-fold higher lateral diffusion for caffeine. The vertical E/D layer shows a saturation of 6 % for caffeine and 8 % for LIP1, which leads to an intra E/D increase of the tissue due to the lateral penetration of 3-fold for caffeine and 1.5-fold for LIP1. A schematic illustration of the lateral movement of caffeine and LIP1 inside the E and D for a FDC setup is shown (Figure 62).

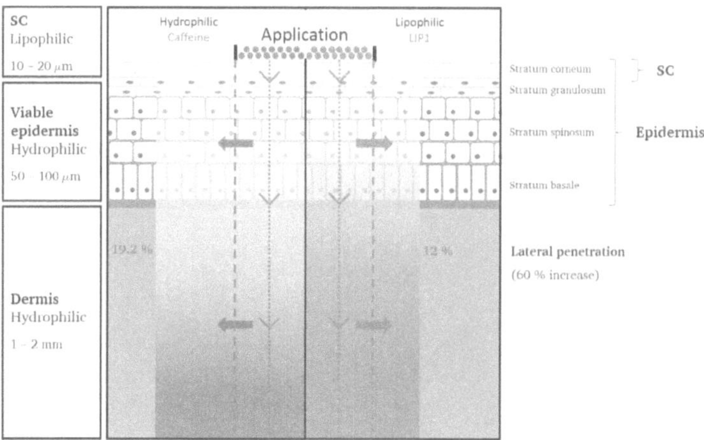

Figure 62: Schematic illustration of the lateral penetration of the FDC setup. Caffeine shows a 60 % increase of the lateral penetration compared to LIP1.

The lateral MD setup shows in Figure 55 for the superficial depth 4.3 % lateral penetration and for the deep depth 7 % lateral movement of caffeine. By looking at the total amount Figure 56 shows no difference between the total lateral penetrated amount, which indicates a conical penetration profile with an increasing lateral movement for an increasing penetration depth. Figure 63 illustrates schematically an increasing conical lateral movement of caffeine and LIP1 for an increasing skin depth. Comparing this with the vertical penetration in Figure 55 the deep implanted membrane should have 35 % less caffeine than the superficial membrane. LIP1 shows in Figure 56 a lateral penetration of 1.3 % for the superficial and 3.3 % for the deep implanted membrane. Figure 57 describes the superficial and deep lateral penetration movement of LIP1 and shows no differences on the penetration amount for the different depth. Comparing this with the vertical penetration in Figure 56 the deep implanted membrane should have 63 % less caffeine than the superficial membrane. The increase of the lateral penetration of the deep implanted membrane, in comparison to a superficial membrane for both actives, indicates an increasing conical profile movement of the active inside the deeper D layer.

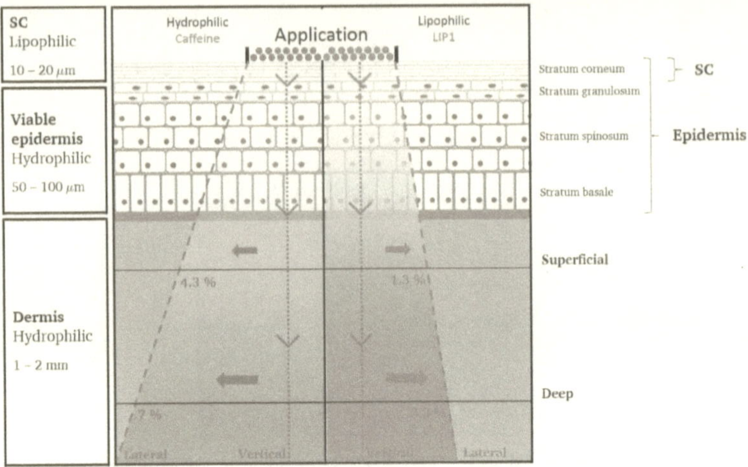

Figure 63: Schematic illustration of the lateral penetration of the MD setup. Caffeine shows a 4.3 % and LIP1 1.3 % increase for lateral penetration for the superficial depth and a 7 % and 3.3 % increase for the deep depth.

The static lateral penetration setup shows in Figure 59 for caffeine a relative lateral diffusion inside the E of 11.3 % and in the D a 22.4% of the recovered amount per E and D layer. LIP1 shows 6.3 % of relative lateral movement inside the E and 0 % of relative lateral movement of the recovered amount of the E and D layer. Comparing this with the amount found per weight of skin layer Figure 60 shows for caffeine a 4 % and 7 % of relative lateral penetration for E and D and for LIP1 2.2 % and 0 %. Overall caffeine showed a higher lateral diffusion characteristic than LIP1, due to the hydrophilic skin environment the lateral movement of LIP1 is reduced as well as the lateral conical movement inside further lateral tissue. Nguyen et al. showed a contribution of lateral movement of actives inside the D as contribution to the overall penetration [78]. Lateral spreading or diffusion was investigated for the SC layer, but a lateral penetration inside this and other layers is still not fully understood [77, 169, 170].

8   Outlook

For a further understanding how the lipophilic characteristics of an API influences the penetration ability, a diverse set of molecules with a log P ranging -2 to 5 need to be tested for their percutaneous penetration ability.

Currently, API efficacy testing as a proof of principle is accomplished via 2D cell culture and distinguishes the minimum concentration needed to achieve a desired biological response. This needs to be cross validated with the penetration ability of the API and if the minimum amount of active ingredient can be located at its target-site. Due to a lack of blood vessels, connection and communication of cells using assays are not predictive for viable skin. The minimally invasive MD penetration system enables a double read out for *ex vivo* skin, with penetration ability controlled via a skin response. The skin reactions are monitored *in situ* on the basis of its metabolites and their utilization as beneficial novel active ingredients.

However, HPLC is not sensitive enough to distinguish the API concentration for beneficial MD analysis and needs to be coupled with mass spectrometry to enable API detection in smaller volumes. This would enable more frequent sampling, leading to more accurate kinetic analysis which would in turn help in the interpretation of data for active ingredients with low penetration ability. Those *in vitro* and *ex vivo* assays give information about the API penetration ability and biological response, but the absence of the cutaneous metabolism, blood circulation and the possible physiological changes which will take place after excision are limiting factors [112].

To overcome this, animal *in vivo* penetration studies of molecules with diverse lipophilic characteristics would provide information regarding target-site penetration and the biological response. But for an understanding of the penetration ability and the desired biological effect of active ingredients inside the human skin, percutaneous human *in vivo* clinical studies are essential. Therefore, the penetration ability of APIs and their biological effect are studied in parallel and the systemic uptake can be determined. This setup has the drawback of no direct correlation between the site of application and API distribution inside the tissue [112]. A combination of *in vitro, ex vivo,* and *in vivo* analyses of artificial membranes, porcine skin, and human skin, provides the most reliable percutaneous penetration data for APIs in terms of biological uptake and function.

The aim of this study was to understand target-site penetration and the influence of lipophilicity on the penetration of the API by analyzing the penetration of hydrophilic caffeine and lipophilic LIP1 in porcine skin. Target-site penetration is of interest to determine the functionality of active compounds and to understand overall skin penetration. Therefore, *in vitro* and *ex vivo* FDC and MD penetration setups were used in combination with the Strat-M® membrane, porcine ear skin, and human skin as surrogate methods for *in vivo* human testing. The selection of the penetration model, skin surrogate, APIs, dosing amount and RF need to be based on the research question being addressed.

Thereby multiple factors influence the penetration ability, to increase the penetration of APIs an optimised vehicle, containing penetration enhancer are used. In combination with a high application volume and API concentration at high penetration temperature the penetration rate is increased and optimised.

Artificial membranes and especially the Strat-M® membrane support penetration studies, and also have the potential of predicting to help elucidate the influences on penetration [161, 162]. Porcine ear skin is a valuable human skin surrogate, providing information on the influences on penetration of APIs, but its higher penetration rate than human skin also needs to be taken into account.

The skin is build up of multiple layers, the SC provides a lipophilic barrier for the penetration of APIs, followed by the hydrophilic E and D environment. Lipophilic APIs like LIP1 show a good penetration potential into the SC, by overcoming this barrier, the penetration rate is reduced by the hydrophilic E and D environment and the active accumulates in those layers, till the kinetic and gradient drive the active into the RF. These results reinforce the observation that the lipophilic SC, and the hydrophilic E and D environment impacts the penetration of active ingredients into skin. The lipophilic environment slows down the penetration rate of hydrophilic molecules and the hydrophilic environment slows down the rate of lipophilic active ingredients.

For caffeine and LIP1, different donors show an intra-donor trend of active ingredient concentration inside the SC, E, and D layer. Each donor shows an intra-donor equilibrium of active ingredient in the SC, E, and D which varies within donors. This equilibrium is layer specific and an entrapped compound reservoir in the different skin layers, along with a layer dependent maximum capacity of a hydrophilic and lipophilic compound is observed. The aqueous skin layers incorporate less lipophilic active ingredient than the lipophilic skin layer for hydrophilic or lipophilic active ingredients.

The lateral penetration studies indicated a conical penetration profile in the presence of increasing lateral movement with increasing penetration depth. The increase in lateral penetration for deep implanted membranes, in comparison to a superficial membrane for both active ingredients, indicates an increasing conical movement of active ingredients inside the deeper D layer. Overall, caffeine showed higher lateral diffusion than LIP1 did; due to the hydrophilic skin environment, the lateral movement of LIP1 is reduced as well as the lateral conical movement further inside lateral tissue.